IET TELECOMMUNICATIONS SERIES 72

Introduction to Digital Wireless Communications

Other volumes in this series:

Introduction to Digital Wireless Communications

Hong-Chuan Yang

The Institution of Engineering and Technology

Published by The Institution of Engineering and Technology, London, United Kingdom

The Institution of Engineering and Technology is registered as a Charity in England & Wales (no. 211014) and Scotland (no. SC038698).

© The Institution of Engineering and Technology 2018

First published 2017

The Institution of Engineering and Technology
Michael Faraday House
Six Hills Way, Stevenage
Herts, SG1 2AY, United Kingdom

www.theiet.org

British Library Cataloguing in Publication Data
A catalogue record for this product is available from the British Library

ISBN 978-1-78561-160-5 (hardback)
ISBN 978-1-78561-161-2 (PDF)

Typeset in India by MPS Limited
Printed in the UK by CPI Group (UK) Ltd, Croydon

To Jie, Jamie, and Jason

Contents

Preface

Motivation and goal

This book serves an efficient introduction to digital wireless transmission technologies in current and future wireless communication systems. The field of wireless communications has witnessed an explosive fast growth since the deployment of the first cellular system in the 1970s. Various advanced-transmission technologies, including multiple-input-multiple-output (MIMO) transmission, multicarrier/orthogonal frequency division multiplexing (OFDM) transmission, channel adaptive transmission, and cooperative relay transmission, have been developed and deployed to meet the growing demand for high data rate wireless services. As a result, digital wireless transmission reaches a new level of maturity. A variety of candidate solutions for reliable and efficient digital information transmission become available for diverse wireless channel conditions. While new generation of wireless systems continue to emerge, the fundamental principle and tradeoff of digital wireless transmissions will remain the same.

This book systematically presents the operating principles and design tradeoffs of various digital wireless transmission technologies. The objective is to provide, in a compact volume, a thorough treatment of these technologies with sufficient theoretical depth that is suitable for college seniors, beginning graduate students, and practicing engineers. The intended audience should have basic training in analog and digital communication systems, at the level of first college course on communication systems. Through innovative illustrations and enlightening examples, this book helps the reader develop an accurate and clear understanding of these transmission technologies. The goal is to build a solid foundation for students/engineers to comfortably deal with future system evolution and technology advances, which enables them to pursue academic research and product development in this fascinating field.

Features

This book grew out of my lectures for a senior elective course on wireless and mobile communications at the University of Victoria. Over the past years, I have been constantly updating the course contents to integrate new developments in the field. In particular, earlier emphasis on power efficient nonlinear modulation and spread spectrum/code division multiple access (CDMA) were gradually evened out later by multicarrier transmission/OFDM, adaptive transmission, and MIMO transmission technologies. The course enjoys continuing popularity among senior students and fresh graduate students, some of whom have not taken the prerequisite course on digital communications. To make the course accessible to students with basic training on communication systems, I redesigned the course by focusing on linear bandpass modulation schemes. Applying the I/Q representation of the bandpass signals, a general digital transmission framework is established for the analysis and design of various wireless technologies. Through this innovative approach, the students can follow advanced wireless transmission technologies while reinforcing their communication theory and system background. The students can also appreciate the applications various mathematical tools, including probability theory, linear algebra, and constraint optimization, and signal processing techniques, such as discrete Fourier transform, zero-forcing, and minimum mean square error (MMSE) processing, in wireless communication system design and analysis. The resulting course becomes as an ideal capstone course for undergraduate students in system option.

This book follows the same general approach. Specifically, the transmission system adopts linear bandpass modulation, which is widely used in advanced wireless communication systems. The reader may refer to many excellent textbooks for more comprehensive treatment of digital modulation schemes. Meanwhile, the presentation of wireless transmission technologies are made rather generic, without being specific to a particular wireless system or standard. Other unique features of the book include the following. Newly presented mathematical models are immediately applied to solve engineering problems, e.g., shadowing/path loss models for outage/coverage analysis. Transmission technologies and analytical tools are introduced as the solutions to engineering design and analysis problems. For example, the statistical fading channel models are developed in response to the performance quantification need for digital transmission over fading channels. Equalization and multicarrier transmission are proposed to mitigate frequency selectivity of the channel. Carefully designed illustrations are developed to facilitate easy understanding of the complex transmission schemes and various design tradeoffs are demonstrated with detailed examples. Finally, this compact volume manages to provide a concise but informative coverage of several advanced transmission technologies, including multiuser diversity transmission, cooperative relay transmission, and multiuser MIMO transmission.

Organization

Chapter 1 provides a brief overview of wireless communications. The primary objective is to establish the point-to-point transmission scenario for later chapters by discussing various radio spectrum access schemes and duplexing schemes.

Chapters 2 and 3 cover wireless channel modeling. Chapter 2 is dedicated to models for large-scale channel effects, i.e., path loss and shadowing. These models are immediately applied to the cochannel interference and coverage analysis of cellular wireless systems.

Chapter 3 studies the characterization of multipath fading effect. After discussing the fading classification by comparing fading channel characteristics and transmitted signal properties, we present simplified models for the popular slow frequency-flat and frequency-selective fading scenarios.

Chapter 4 considers digital transmissions over flat fading channels. Linear band-pass modulation schemes are first introduced, followed by the study of the effects of complex flat fading channel gain on their detection performance. Countermeasures to channel phase, including channel estimation and differential modulation, are explained in details. The statistical characterization of channel amplitude is developed and applied to the quantification of performance degradation due to fading.

Chapter 5 covers one of the most effective fading mitigation techniques, diversity combining. Several fundamental combining schemes are presented for reception diversity scenario. The tradeoff between performance and complexity offered by different schemes are quantitatively investigated. The transmit diversity solutions with and without channel state information at the transmitter are also explained in details.

Chapter 6 shifts the focus to selective fading channels. After explaining the negative effects of selective fading, two classes of transmission technologies, namely equalization and multicarrier transmission, are presented as countermeasures. The discrete implementation of multicarrier transmission, the OFDM technology, is introduced in particular.

Chapter 7 considers spread-spectrum transmission technologies. Both direct sequence and frequency-hopping spread-spectrum solutions are studied. The underlying mechanisms that lead to those desirable features of spread-spectrum transmission are explained and demonstrated. The concept of RAKE receiver and the CDMA scheme are presented and analytically investigated.

Chapter 8 is dedicated to capacity and coding for fading wireless channels. The capacity limits of wireless channels together with capacity-achieving transmission strategies are introduced. Sample coding schemes that can approach capacity over additive white Gaussian noise (AWGN) channels are presented. The chapter is concluded with the investigation of interleaving technique as an effective coding solution for fading channels.

Noting that a selective fading channel can be converted into parallel flat fading channels with multicarrier/OFDM technology, Chapter 9 shifts gear back to flat fading channels by investigating channel adaptive transmission technology. Various adaptation strategies are presented and analytically studied, together with practical implementation issues associated channel adaptation.

Chapter 10 covers multiple-antenna transmission and reception, i.e., MIMO technology. After demonstrating the diversity and capacity potential of MIMO channels, several transmission strategies that can exploit diversity gain or multiplexing gain or both from MIMO channels are studied. The diversity multiplexing tradeoff is also characterized.

Chapter 11 presents several advanced wireless transmission technologies, including multiuser diversity transmission, cooperative relay transmission, and multiuser

MIMO transmission. These technologies improve the performance and efficiency of wireless transmission by in one way or another exploring multiple antennas at different mobile terminals.

The appendix covers four different mathematical tools used in the text. Appendix A.1 discusses the Fourier transforms for both continuous-time and discrete-time analysis. Appendix A.2 covers the basic concepts of probability, random variables and random processes. Appendix A.3 deals with vectors and matrices, including their properties and operations. Finally, Appendix A.4 presents the Lagrange multiplier method as a useful optimization tool.

Usage

The book is designed to be a textbook for senior undergraduate or first year graduate course on wireless communications. The required background would be a undergraduate course on communication systems. Certain experience in digital signal processing, probability and random processes, linear algebra is assumed. Essential materials on these topics are summarized in the Appendices. Background in digital communications is helpful but not required.

The majority of materials in the book can be covered in a quarter or semester course. Chapters 1–6 contain the core materials. The remaining chapter may be covered selectively. Depending on the students' background and specific curriculum structure, certain chapter/section can be omitted without affecting the flow. Examples include equalization in Chapter 6, frequency hopping in Chapter 7, coding in Chapter 8 and the material in Chapter 11.

The book will also serve the need of engineers and postgraduate researchers by providing a succinct technical treatment of digital wireless transmission technologies. These materials are presented without assuming extensive digital communications background from the readers. Special emphasis will be placed on the important tradeoff of performance versus complexity throughout the presentation. The design principles of different technologies are illustrated with carefully designed figures and problem-solving examples. The readers can also check their own understanding with extra practice problems at the end of each chapter.

Acknowledgments

I would like to thank my former teachers in China and professors at the University of Minnesota for enlightening me from different aspects. I could not reach my current stage of career without the foundation that they helped me build over years. I need to thank two people in particular for all my professional endeavors. *Karen K. Yin* of the University of Minnesota gave me the initial opportunity to experience learning and research in north America. Her caring supervision and constant support enabled me to achieve more than I expected and continue to serve as a source of inspiration and motivation for me. Mohamed-Slim Alouini introduced me to the fascinating field of wireless communication research. With his knowledge, passion, and kindness, he is an ideal role model as well as an excellent PhD advisor. I am deeply indebted to his continuous encouragement, inspiration, and friendship.

I am grateful to my collaborators over the years, who have enriched my knowledge and experience in wireless communications. I had the honor to have collaborated with talented researchers from Canada, China, Germany, Korea, Norway, Qatar, Saudi Arabia, and United States. My collaborators include, but not limited to, Seyeong Choi, Pingyi Fan, Mazen O. Hasna, Young-Chai Ko, Khalid B. Lataief, Andrea Mueller, Sungsik Nam, Geir E. Øien, Khalid A. Qaraqe, Kamel Tourki, and Ke Xiong. I am also indebted to many colleagues and friends, who have provided valuable support to my career. I am especially grateful to George G. Yin, Ahmed Tewfik, Vijay Bhargava, Sofiène Affes, Pooi Yuen Kam, Nikos Sidiropoulos, Victor Leung, Li-Chun Wang, Robert Schober, Chintha Tellambura, Marco Chiani, Geoffrey Y. Li, Ross Murch, Norm Beaulieu, Ying-Chang Liang, and Georgios B. Giannakis.

I am indebted to the wonderful colleagues and students at the University of Victoria. The dynamic, stimulating, and friendly environment greatly helped foster this book. My appreciation goes particularly to Pan Agathoklis, Ashoka Bhat, Jens Bornemann, Thomas Darcie, Nikitas Dimopoulos, Peter Driessen, Fayez Gebali, Aaron Gulliver, Wu-Sheng Lu, Thomas Tiedje, and Kui Wu. I want to thank my former and current graduate students. Through the interaction with them, I was able to better polish my understanding of important wireless concepts and transform some of them in a more accessible form into this book. I am privileged to teach a diverse student body on various subjects, which motivates me to understand things thoroughly and express them more simply. This process helped improve my teaching, my research, and myself as a person.

I want to thank Valerie Moliere, the commissioning editor at IET, whose remarkable talent has identified the potential of this project. The valuable support from Mr. Paul Deards, Ms. Jennifer Grace, and Ms. Olivia Wilkins at IET during the preparation of the book are much appreciated.

I am deeply grateful to my family for their love, support, and encouragement. I believe I inherit the engineering thinking from my father and the educator passion from my mother. I would not consider such an enduring project without the support of my wife, Jie. She is the best wife, mother, and friend I could ever imagine. My lovely kids Jamie and Jason are great supporting fans of the project. They try to help in whatever ways they can, including checking progress, discussing word choices, and even correcting typos sometimes. This book is dedicated to these three most important people in my life.

Hong-Chuan Yang
Victoria, Canada

Chapter 1
Introduction to wireless transmission

Wireless communications is a fascinating field with extensive scope. This book primarily focuses on the point-to-point wireless transmission technologies. In this introductory chapter, we first present a brief overview of wireless communications. Then the access of wireless communication medium, the radio spectrum, is discussed, where sample schemes for both multiple access and random access are introduced. After that, we discuss the available duplexing schemes for two-way wireless communication systems. With the point-to-point transmission scenario thus established, the chapter is concluded with an outline of digital wireless transmission technologies.

1.1 Overview of wireless communications

Wireless communications generally refer to any form of information transmission from one terminal to another without physical wire connection between them. Any form of information delivery over the air falls into the regime of wireless communications. Early examples of wireless communications include smoke signals and flag signals in the preindustrial age. Sign language serves as a modern example. All of these sample wireless communication systems rely on the existence of a line of sight between two participating terminals. Such requirement for direct line of sight becomes unnecessary when information are electronically transmitted using electromagnetic signals.

In the modern era, the term 'wireless communication' commonly relates to the electronic information transmission by the means of electromagnetic wave propagation. The information is always first converted into proper electronic format before being transmitted. Sample wireless communication systems include radio broadcasting, TV broadcasting, satellite communications, cellular telephone, wireless local area network (WLAN), etc. Early wireless communication systems used analog transmission technologies. With the successful application of digital wireless transmission technologies in early 1990, the wireless industry experienced an explosive growth period and is still the fastest growing segment of the information communication technology (ICT) sector.

Wireless transmission relies on the physical phenomenon of electromagnetic wave propagation. When the current in a conducting device is varied, a varying magnetic field is generated, which in term leads to a changing electrical field. The resulting electromagnetic wave can be used to carry information if we vary the current of the conducting device according to the information to be transmitted, which is commonly referred to as the modulation process. The information may be recovered at the receiving end by tracking the variation pattern of the incident electromagnetic wave, also known as the demodulation process. These processes form the physics fundamentals of electronic wireless communication.

The first wireless transmission based on electromagnetic wave propagation was carried out by Marconi in 1895. The amplitude of the radiated wave was modified to carry information, which leads to the well-known amplitude modulation (AM) systems. Most early radio broadcasting used AM. In 1935, Armstrong developed a new radio transmission scheme, where the frequency of the electrical field variation is modified according to the information to be transmitted. With its superior performance over the AM counterpart, the resulting frequency modulation (FM) system gain increasing acceptance. Both AM and FM systems are still widely used today in radio broadcasting and analog TV broadcasting.

The cellular mobile radio systems are one of the most important developments in the wireless industry. Built upon the cellular concept originated from AT&T Bell laboratory in the late 1960s, these systems can deliver two-way voice communication services to mobile terminals over a wide area. The first cellular system was deployed in late 1970, using FM as the transmission scheme, which led to huge commercial success. Wireless transmission also finds applications in data packet transmission between computers. The first packet radio network was developed at the University of Hawaii and allowed computers on different islands to communicate with a central computer. Such early wireless data networks for both wide areas and local areas were not commercially successful due to their low data rate and poor coverage. In fact, wide area wireless data networks was eventually superseded by the data capability of digital cellular systems.

The commercial success of cellular systems in the late 1980s stimulated intense interest in the development of efficient wireless transmission schemes for better quality and higher spectral efficiency. The basic design objective in the 1990s is to accommodate as much voice channels as possible with the given radio spectrum. As a result, most cellular standards started to adopt digital transmission schemes. Meanwhile, digital transmission technologies are also introduced to wireless local area data networks, to narrow the performance gap between wired local area networks (LANs) and WLANs. The digitized version of WLANs, also known as WiFi, demonstrated to be another enormous commercial success. The newer generations of WLANs based on the IEEE 802.11 family standards become the preferred data communication method for laptop computers and other portable devices owing to their convenient deployment and easy usage.

With the explosive growth of multimedia resources on the Internet, newer generations of wireless communication systems strive to support high data rate Internet access. Such paradigm shift stimulates intense development of wireless transmission technologies. Advanced coding and modulation schemes are deployed to improve spectral efficiency. Various performance improving technologies are developed to achieve better reliability. Transceiver design for various operating environment are proposed to enhance coverage. These developments bring digital wireless transmission technologies to a new level of maturity. A variety of candidate solutions become available for the reliable and efficient transmission of digital information over diverse wireless channels. The following chapters provide a comprehensive introduction to these transmission technologies.

1.2 Spectrum access

Wireless systems use radio frequency resource to provide tetherless communication services. The radio frequency bands available for wireless communications span from several kHz to as high as hundreds of GHz. The allocation and control of the radio spectrum in each county is handled by appropriate government agencies, Federal Communications Commission (FCC) in the United States as an example. Internationally, the spectrum allocation is coordinated by the International Telecommunications Union (ITU). The frequency bands and their sample usage are summarized in Table 1.1. Certainly spectrum are restricted to military usage or other special purposes, while others for commercial usage. Most commercial spectrum are licensed. Service providers can purchase a license to gain the exclusive right for using a portion of such spectrum. As an example, cellular phone systems operate over licensed spectrum around 900 MHz and 2 GHz. Certain portions of spectrum are free to use without a license. These unlicensed bands, including the Industrial, Scientific, and Medical (ISM) bands and unlicensed National Information Infrastructure (U-NII) bands, are set aside to encourage innovation and facilitate low-cost applications. In particular, WiFi and bluetooth applications use the unlicensed bands around 2.4 and 5.5 GHz. The spectrum allocation of selected commercial wireless systems in North America are listed in Table 1.2.

Wireless systems typically serve multiple users with the allocated spectrum. As such, the system must effectively coordinate the spectrum access of eligible users for their communication needs. The same spectrum can be reused over geographically separated areas, provided that the separation is sufficient large such that mutual interference between transmissions is negligible. For example, the same frequency band

Table 1.1   Radio frequency bands and their sample usages

Frequency bands	Name	Sample usages
3–30 kHz	Very low frequency (VLF)	Sonar and navigation
30–300 kHz	Low frequency (LF)	Navigational aides
300–3,000 kHz	Medium frequency (MF)	Maritime and AM radio
3–30 MHz	High frequency (HF)	Aircraft and telegraph
30–300 MHz	Very high frequency (VHF)	VHF TV and FM radio
300–3,000 MHz	Ultra high frequency (UHF)	Surveillance and land mobile
3–30 GHz	Super high frequency (SHF)	Microwave and satellite
30–300 GHz	Extremely high frequency (EHF)	Railroad and experiments

Table 1.2   Spectrum allocation of sample wireless applications

Applications	Band	Frequency
AM radio	MF	535–1,605 kHz
FM radio	VHF	88–108 MHz
Broadcast TV	VHF	174–216 MHz
	UHF	420–890 MHz
Cellular phones	UHF	806–902 MHz, 1.85–1.99 GHz
Cordless phones	ISM band I in UHF	902–928 MHz
Bluetooth	ISM band II in UHF	2.4–2.4835 GHz
Wireless LAN	ISM band II in UHF	2.4–2.4835 GHz
	U-NII bands in SHF	5.15–5.825 GHz
Satellite TV	SHF	12.2–12.7 GHz
Fixed wireless	EHF	38.6–40 GHz

is used in different metropolitan areas to provide AM and FM radio services. When the eligible users are in the same area, their spectrum access is handled by proper access schemes. In general, there are two types of spectrum access schemes, namely *multiple access* and *random access*.

1.2.1   Multiple access

Multiple access schemes divide the spectrum resource into channels and allocate each eligible user with a dedicated channel on a first-come-first-serve basis. Such schemes are suitable for applications with continuous transmission demand, such as voice and video services. Depending upon how the spectrum resource is divided into user channels, we arrive at several multiple access schemes. Frequency-division multiple access (FDMA) is the most fundamental multiple access scheme. With FDMA, each user channel corresponds to a dedicated frequency band, as illustrated in Figure 1.1(a). Typically, there are guard bands between adjacent frequency channels to account for imperfect filtering and mitigate interchannel interference. Almost all analog wireless systems use FDMA scheme.

Time-division multiple access (TDMA) scheme becomes feasible with the introduction of digital transmission technologies. With TDMA, each user channel corresponds to a sequence of cyclically repeated time slots over a particular frequency band, as shown in Figure 1.1(b). Specifically, a user will periodically access the frequency band for a short duration. The duration of access is usually much shorter than the accessing period. As such, digital compression technology is required to support voice or video services. Several second generation cellular systems use TDMA as their multiple access scheme.

Code-division multiple access (CDMA) is an access scheme based on spread-spectrum transmission technology. With CDMA, each user channel corresponds to a spreading code applied over a particular frequency band, as illustrated in Figure 1.1(c). Different users will transmit over the frequency band simultaneously while using different spreading codes. The receiver uses the orthogonality property of the codes

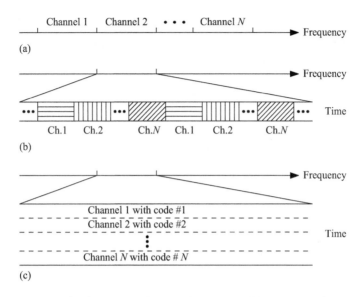

Figure 1.1 Multiple access schemes: (a) FDMA, (b) TDMA, and (c) CDMA

to separate user transmission. Interuser interference may occur due to non-orthogonal codes, synchronization error, and/or the channel effects. CDMA is a popular multiple access scheme in third generation cellular systems.

1.2.2 Random access

When the transmission demands are bursty, as the case in most data applications, random access schemes become more efficient. With random access, the user information is usually divided into packets of a certain number of bits, which may also contain error detection/correction bits and control bits. Multiple users attempt to transmit their own packets over the same channel without overhead signaling to separate their transmission. As such, users' transmission may collide with one another, in which case all transmission may be unsuccessfully. Packet error may also occur due to noise or other channel impairments. Typically, the packets received in error will be retransmitted, which increases the traffic load.

Pure ALOHA is the most fundamental random access scheme. With pure ALOHA, users will transmit their packets whenever a packet becomes available. While simple, the performance of pure ALOHA is poor, mainly due to the frequent packet collision. As illustrated in Figure 1.2(a), packet transmission of two users may partially overlap, rendering both packet transmission unsuccessful. The performance of pure ALOHA can be considerably improved when the time axis is slotted and packet transmission, if any, always starts at the beginning of each slot. Packet transmissions with the resulting slotted ALOHA scheme will not partially overlap at all, as shown in Figure 1.2(b). The reduction of packet collision and performance improvement with

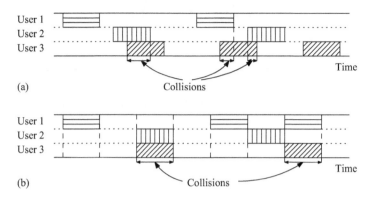

Figure 1.2   Random access schemes: (a) pure ALOHA and (b) slotted ALOHA schemes

slotted ALOHA comes at the cost of synchronization requirement on all users in the system.

Packet collision and the subsequent packet retransmission have a detrimental effect on the performance of random access schemes. The effect of collision can be mitigated with carrier sensing, collision detection, and collision avoidance. Carrier sensing multiple access (CSMA) is a random access scheme with improved performance over ALOHA schemes. With CSMA, a user with packet to send will first sense the channel and transmit only if no ongoing transmission is sensed. If ongoing transmission is sensed, the user will back off for a random time period and sense the channel again. It may happen that two users sense the channel free and start their transmission at the same time, leading to collision. There are generally two variants of CSMA depending on how to address these residual collision. With CSMA collision detection (CSMA/CD) scheme, the user will detect if another user transmission starts at the same time or not. If a collision is detected, the user will stop transmission and wait for a random time period before restarting the process. Since it is very challenging for wireless terminals to listen while transmitting, another variant of CSMA, CSMA with collision avoidance (CSMA/CA), was developed. With CSMA/CA, the user, after sensing channel free, sends a *request-to-send* packet to inform the users in the area, including the receiver, about its upcoming transmission. Then the receiver will send a *clear-to-send* packet to signal the user to start data transmission. As a result, the subsequent data transmission is highly unlikely to experience collision.

1.3   Duplexing

The huge commercial success of wireless systems originates from their capability of "wirelessly" providing the same communication services as their wired counterpart. Communication services can be generally divided into two types based on the direction of information flow, i.e., one-way communication, e.g., AM/FM

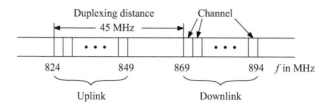

Figure 1.3   Frequency division duplexing

radio and TV broadcasting services, and two-way communication, e.g., telephone and video-conference services. Two-way communication services are further divided into half-duplex two-way and full-duplex two-way, depending whether two terminals can send information simultaneous or not. Cellular wireless systems were originally designed with providing full-duplex two-way voice communication services as the main objective.

Wireless communication systems typically achieve full duplex service by explicitly identifying the channel in each communication direction. Considering the cellular system example, the system needs to assign a certain channel for uplink transmission from the mobile to the cellular base station (BS) and another channel for downlink transmission from the BS to the mobile. Depending how these channels for different communication directions are separated, we have two possible duplexing schemes, namely frequency division duplexing (FDD) and time division duplexing (TDD). Specifically, FDD systems use different frequency range for each communication direction, whereas TDD systems use different time slots for each direction.

Let us illustrate these two duplexing schemes using a cellular system example. Most first and second generation of cellular systems use FDD. Figure 1.3 shows the duplexing configuration for a sample second generation system. The uplink channels use the frequency band of 824–849 MHz, whereas the downlink channels use the frequency band of 869–894 MHz.[1] The pairs of uplink/downlink channels are separated by a gap of 45 MHz, which is referred to as *duplexing distance*. While such duplexing distance facilitates the transceiver design in terms of minimizing cross-talk between transmitted and received signal, strictly abiding to this required gap may leads to a certain waste of radio spectrum when the allocated spectrum is not exact even multiple of such gaps. Note from Figure 1.3 that the spectrum of 849–869 MHz cannot be used with FDD unless the system operator also owns the spectrum of 894–914 MHz.

TDD scheme exhibits more flexibility in spectrum exploration. With TDD, both uplink and downlink channels will be using the same frequency band, but access it in at different time slots. Specifically, as illustrated in Figure 1.4, the access of the common frequency bandwidth is slotted in time domain. Periodically repeated time

---

[1]For the reason to be discussed later in Chapter 2, the lower frequency range is allocated to uplink transmission to save the transmission power of battery-powered mobile terminals.

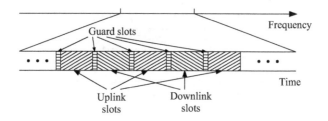

Figure 1.4   Time division duplexing

slots are used for uplink and downlink transmissions respectively. Short guard slots are usually inserted between uplink and downlink slots to account for propagation delay and transceiver reconfiguration. While appearing like a half-duplex implementation of two-way communication, TDD scheme enables full-duplex experience with the application of digital signal processing and compression technologies. Essentially, information to be transmitted in each direction is digitized and compressed to fit into the allocated time slots.

TDD scheme also enjoys several subtle desirable features for advanced digital transmission. Since both uplink and downlink transmission use common frequency bandwidth, the channel characteristics of two communication directions may be very similar, which is usually referred to as the *channel reciprocity* property. With channel reciprocity, the transmitter may be able to estimate the forward link channel quality based on the received signal over reverse link. Therefore, TDD facilitates the access of channel state information at the transmitter side, which can be explored to improve the transmission quality and efficiency. Note that with FDD scheme, the transmitter can only obtain the forward link channel state information through the explicit feedback from the receiver.

Another desirable property of TDD scheme is the reconfigurability to support unbalanced traffics. The duration of uplink and downlink time slots can be adjusted based on the traffic load. For voice communication, the traffic load in both directions should be similar, which leads to equal slot duration for uplink and downlink. For data and multimedia communications, the downlink load may be larger than the uplink load, which can be accommodated by decreasing uplink slot duration and increasing downlink slot duration. Such adjustment can be made on the fly with proper reconfiguration of the transmitter and receiver. WiFi systems, bluetooth systems, and selected cellular systems use TDD scheme.

Example: Multiple access and duplexing

GSM is a popular second generation cellular system. A version of GSM system use the spectrum band of 890–960 MHz to offer full-duplex voice communication service. The system adopts FDD scheme with duplexing distance of 45 MHz. The multiple access scheme is a combination of FDMA and TDMA, where the spectrum for each transmission direction is divided into frequency subband of

200 kHz and each of the subband carries 8 voice channels in a time division fashion. Determine the total number of full-duplex voice communication session that the system can simultaneously support with the available spectrum. What if the system adopts TDD scheme with frequency subbands of 200 kHz, each of which carries four full-duplex voice communication sessions?

Solution: To satisfy the duplexing distance constraint, the frequency subband for uplink transmission should be separated from that for downlink transmission by 45 MHz. As such, the downlink subband corresponding to the first uplink subband of 890–890.2 MHz will be 935–935.2 MHz. Similarly, the uplink subband corresponding to the last downlink subband of 959.8–960 MHz will be 914.8–915 MHz. Therefore, with FDD implementation, the system can only use the frequency band 890–915 MHz for uplink transmission and 935–960 MHz for downlink transmission. The frequency band of 915–935 MHz cannot be used. The total number of full-duplex channel is $8 \times (915 - 890)/0.2 = 1,000$.

If TDD scheme is used, the number of full-duplex session that the system can support becomes four times the number of 200 kHz subbands that the spectrum can accommodate, i.e., $4 \times (960 - 890)/0.2 = 1,400$.

1.4 Digital wireless transmissions

The structure of a basic digital wireless transmission system is shown in Figure 1.5. The information generated by the source first goes through the source encoder, which removes the redundancy of the source information as well as performs analog to digital conversion, if necessary. The channel encoder then added some controlled redundant bits to provide error detection and correction capability. The resulting coded information, in the form of binary bit sequences, are then processed by the modulator. The objective of the modulator is to convert the coded binary information into a format suitable for wireless transmission. The modulated signal, usually in the form of sinusoidal, is processed by RF circuits and finally transmitted from the antenna.

The transmitted signal will propagate through wireless channels, which introduce attenuation, distortion, interference, as well as noise. The signal collected by the received antenna will be first processed by proper RF circuits, which performs filtering and down-conversion among other operations. Then the signal is demodulated and decoded to recover the source information. Compared to the analog transmission systems, digital wireless transmission has many advantages, including higher spectral efficiency, resistance to channel impairments, improved security and privacy, and error detection/correction capability. As such, most current and emerging wireless systems are adopting digital transmission schemes.

Early development of digital wireless transmission schemes was primarily concentrating on the modulation scheme and coding scheme design. Power efficient

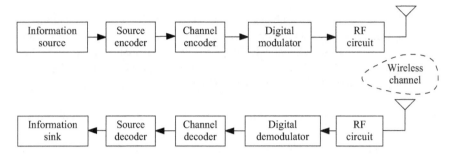

Figure 1.5   Digital wireless transmission system

nonlinear amplifier mandates modulation schemes that generate modulated signals with small envelop variation. As such, $\pi/4$-DQPSK (differential quadrature phase shift keying) and Gaussian minimum shift keying (GMSK) were developed and applied in second generation cellular standards. Both schemes allow for noncoherent detection, which reduces receiver complexity. When the demand for higher spectral efficiency overshadows power efficiency, higher order linear modulation schemes achieving high data rates become more favorable. While requiring coherent reception, linear amplitude/phase modulation facilitates channel adaptive transmission.

The wireless channel introduces fading effect, which can seriously degrade the transmission performance if not properly mitigated. Digital wireless system can adopt classical fading mitigation techniques, such as diversity combining and channel coding, to improve the transmission reliability. Deploying multiple antennas at the receiver or the transmitter demonstrates to be an attractive diversity solution. When multiple antennas are available at both the transmitter and the receiver, i.e., MIMO scenario, the system can achieve both diversity benefit and higher spectral efficiency. When the transmitter has access to the channel quality, adaptive transmission can be implemented to improve transmission efficiency while meeting certain reliability requirement.

The fading wireless channels may also introduce filtering effect and severely distort the transmitted signal. Such frequency-selective fading occurs more frequently for high data rate transmission. Equalization represents a large class of signal processing techniques that digital wireless system can employ to revert such channel effect, as well as other channel imperilments. Spread-spectrum transmission and multicarrier transmission technologies are specifically designed for transmission over selective fading. Spread-spectrum transmission intentionally introduce frequency selectivity for other advantages, such as interference mitigation and path diversity. Multicarrier transmission converts selective fading channel into parallel flat fading channels and has gain wide acceptance with its efficient discrete implementation, i.e., orthogonal frequency division multiple (OFDM). Spread-spectrum and multicarrier transmission technologies also forms the foundation of new multiple access schemes for wireless systems.

1.5 Further readings

Further details about the history of wireless communications can be found in [1, Chapter 1]. The evolution of cellular radio systems are discussed in details in [2, 3]. Reference [4] is specifically structured according to different multiple access schemes. Further details about random access schemes are presented in [5, Chapter 6]. Our presentation is limited to linear modulation schemes. Interested readers may refer to [2, Chapter 6] for the basic principle of GMSK modulation schemes.

Problems

1. Wideband CDMA (WCDMA) is a third generation cellular system using FDD scheme. Assume that the spectrum of 1,885–2,200 MHz is allocated to a WCDMA system. With CDMA implementation, as many as 80 users can simultaneously transmit over a bandwidth of 5 MHz. Determine the maximum number of two-way user channels that the system can support if the required duplexing distance is 175 MHz?
2. The spectrum of 1,698–1,746 MHz is available for an operator to offer broadband wireless service.
   (i) If the operator adopts FDD with one-way channel bandwidth 2 MHz and duplexing distance 30 MHz, how many two-way channels that the spectrum can support?
   (ii) What if the operator adopts TDD with two-way channel bandwidth equal to 4.8 MHz?

Bibliography

[1] A. Goldsmith, *Wireless Communications*, New York, NY: Cambridge University Press, 2005.
[2] T. S. Rappaport, *Wireless Communications: Principle and Practice*, 2nd ed. Upper Saddle River, NJ: Prentice Hall, 2002.
[3] G. L. Stüber, *Principles of Mobile Communications*, 2nd ed. Norwell, MA: Kluwer Academic Publishers, 2000.
[4] S. Haykin and M. Moher, *Modern Wireless Communications*, Englewood Cliffs, NJ: Prentice-Hall, 2005.
[5] D. P. Agrawal and Q.-A. Zeng, *Introduction to Wireless & Mobile Systems*, 4th ed. Boston, MA: Cengage Learning, 2016.

Chapter 2

Large-scale channel effects

The most distinguishing characteristic of wireless communication systems is the wireless channel, which has important implications on the design and analysis of wireless transmission technologies. In this chapter, we first provide an overview of wireless channel modeling. Then, the popular models for large-scale channel effects, i.e., path loss and shadowing, are presented. After that, path loss models and shadowing model are respectively applied to the cochannel interference analysis and outage/coverage analysis of wireless communication systems. The fundamental design principle of cellular system is also presented. The small-scale channel effects will be considered in the next chapter.

2.1 Wireless communication channels

Wireless channel originates from the physical phenomenon of electromagnetic wave propagation. The existence of wave propagation over radio frequency was first predicted by Maxwell and then experimentally verified by Hertz in late nineteenth century. Specifically, any varying electrical or magnetic field will generate electromagnetic wave. Such wave will propagate through several mechanisms, including direct line of sight (LOS) propagation, reflection by ground, wall, and large terrain objects, diffraction around sharp edges, and scattering when hitting small spaces and particles, as illustrated in Figure 2.1. In general, the magnitude of the wave will attenuate as propagation distance increases. The attenuation will be relatively slower along the LOS path, as intuitively expected. While leading to faster attenuation, other propagation mechanisms help extend the propagation beyond LOS, which is highly

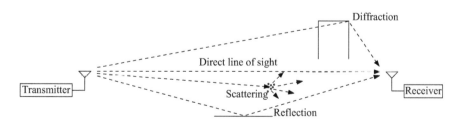

Figure 2.1　Propagation mechanisms of radio wave

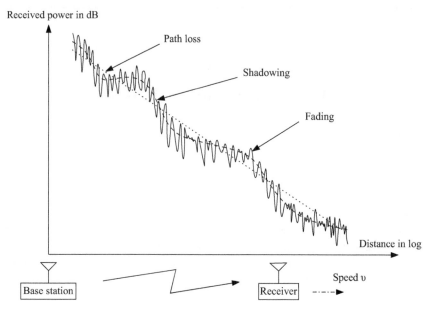

Figure 2.2   The attenuation of received signal power as distance increases

desirable for mobile radio applications. The general consequence is that the radio wave can arrive at the receiver in a wide range of scenarios, while multiple propagation paths between the transmitter and receiver typically exist. Wireless channel is created when the source tries to send some information to a receiver by varying the properties of the radio wave. In this and following chapter, we will discuss the properties and modeling of wireless channels, which have significant impacts on the performance and design of digital wireless transmission schemes presented in the later chapters.

Maxwell has developed a set of differential equations to mathematically characterize radio wave propagation. The complex propagation environment and the unpredictable nature of user mobility make the modeling of wireless channels using Maxwell equations very challenging, if not impossible. To avoid the modeling complication associated with the details of individual propagation paths, the wireless channels are usually characterized by three major effects: (i) *path loss*, i.e., the general trend of power dissipation as propagation distance increases; (ii) *shadowing*, e.g., the effects of large objects, such as buildings and trees, along propagation paths; and (iii) *fading*, resulting from the random superposition of signals received from different propagation paths at the receiver. The received signal power variation due to these three effects is demonstrated in Figure 2.2. Specifically, the path loss effect determines the average received signal power at a distance from the transmitter. Due to the existence or nonexistence of large propagation objects along the propagation path, the received signal power at the same distance from the transmitter may be different, due to the shadowing effect. Therefore, path loss effect determines the average for the shadowing process. The received signal power over a local area that experiences the same level of shadowing may also vary due to the fading effect. Shadowing and path loss effects jointly determine the average of the fading process.

In general, path loss and shadowing are called large-scale propagation effects, whereas fading is referred to as the small-scale effect, as fading effects manifest itself in much smaller time scale. Typically, interference analysis and coverage planning are based on the large-scale effects. The transmission schemes are designed based on the small-scale fading effects. In this chapter, we focus on the models of large scale effects and their applications. The fading effect will be discussed in the next chapter.

2.2 Path loss models

Path loss characterizes the attenuation of signal power as the propagation distance increases. Path loss models typically ignore the variation due to location specific factors and multipath effects, which will be taken into consideration later in shadowing and fading models, respectively. The linear *path loss* is defined as the ratio of the transmitted signal power P_t over the received signal power P_r, i.e.,

$$PL = P_t/P_r. \tag{2.1}$$

After taking logarithm with base 10, we obtain the dB value of path loss, given by

$$PL \text{ dB} = 10 \log_{10} (P_t/P_r). \tag{2.2}$$

Various path loss models have been developed in the literature and used in practice to predict path loss at a certain distance from the transmitter.

A simple, but not necessarily accurate, method to predict the path loss at a certain distance is to use the well-known Friis equation for free space, leading to the *free space path loss* model. With Friis equation, we can calculate the received signal power in a free space environment as

$$P_r = \frac{P_t G_t G_r \lambda^2}{(4\pi)^2 d^2}, \tag{2.3}$$

where P_t is the transmission power, λ is the carrier wavelength, d is the distance between the transmitter and the receiver, G_t and G_r are the gains of the transmit and receive antennas. Applying the definition of path loss, we can predict the path loss as

$$PL = \frac{(4\pi)^2 d^2}{G_t G_r \lambda^2}. \tag{2.4}$$

Therefore, the path loss in free space environment is proportional to the square of propagation distance and the square of the carrier frequency, denoted by f_c, which is related to λ as $f_c = c/\lambda$. Here c denotes the speed of light, equal to 3×10^8 m/s. In dB scale, the path loss is given by

$$PL \text{ dB} = 21.98 - 10 \log_{10} (G_t G_r) - 20 \log_{10} (\lambda) + 20 \log_{10} (d). \tag{2.5}$$

Example: Free space path loss model

Consider a wireless transmission system with carrier frequency of 900 MHz. The gains of the transmit and receive antennas are equal to 1, corresponding to omnidirectional antennas. If the minimum required received signal power at a distance of 1 km from the transmitter is 5 μW, determine the smallest transmission power that can be used with the free space path loss model. What if the carrier frequency is changed to 2 GHz?

Solutions: The   carrier   wavelength   at   900 MHz   frequency   range   is $\lambda = c/f_c = 1/3$ m. Based on the free space model, the path loss at distance 1 km is calculated as

$$PL \text{ dB} = 10 \log_{10}((4\pi)^2) - 20 \log_{10}(1/3) + 20 \log_{10}(1{,}000) = 91.5 \text{ dB.} \quad (2.6)$$

The minimum transmission power level should be higher than the required receive power by 91.5 dB. As such, the minimum transmit power should be

$$P_t \geq P_r + 91.5 \text{ dB} = -25.2 \text{ dBm} + 91.5 = 66.2 \text{ dBm} = 36.2 \text{ dBW.} \quad (2.7)$$

If we repeat the calculation for $f_c = 5$ GHz, which leads to a wavelength of $\lambda = 0.06$ m, the path loss at distance 1 km becomes

$$PL \text{ dB} = 10 \log_{10}((4\pi)^2) - 20 \log_{10}(0.06) + 20 \log_{10}(1{,}000) = 106.4 \text{ dB.} \quad (2.8)$$

The minimum transmission power becomes 41.1 dBW or 12.88 kW. We can clearly observe the effect of faster path loss with higher carrier frequency.

Since Friis equation is derived for the free space environment, where only the direct LOS path from the transmitter and the receiver exists, the above free space path loss model is suitable to predict path loss over LOS path. In real-world environment, there will be multiple propagation paths between the transmitter and the receiver, as the result of other propagation mechanisms. At least, there should be a ground reflection path, as illustrated in Figure 2.3. The resulting two-ray propagation model applies to wireless transmission over rural area. Both measurement and theoretical analysis has verified that the received signal power level will no longer be monotonically decreasing function of the distance due to the addition of ground reflection path. In fact, the signal from two paths may cancel each other at certain distances. Therefore,

Figure 2.3   Two-ray propagation scenario

the free space path loss model has limited applicability. One approach to generalize the free space model is to trace the receive signal power along different propagation paths and consolidate them together at the receiver, resulting the so-called ray-tracing model. While much more accurate than free space or other models, the ray-tracing model requires location-specific propagation details, which may not be feasible in time varying mobile environment.

Another approach is to use measurement data to build empirical path loss model. The most famous empirical model is the Okumura–Hata model, which was built by Hata based on the data collected by Okumura in the city of Tokyo, Japan. The initial model applies to the frequency range of 150–1,500 MHz and involves a set of formulas to predict the median path loss in different operating environment. For example, the median path loss for urban area, is calculated as

$$PL_{\text{median}} \text{dB} = 69.55 + 26.16 \log_{10}(f_c) - 13.82 \log_{10}(h_t)$$
$$-a(h_r) + (44.9 - 6.55 \log_{10}(h_t)) \log_{10}(d), \tag{2.9}$$

where h_t is the base station antenna height, h_r is the mobile antenna height, and $a(h_r)$ is the correction factor for mobile antenna height. In larger cities with $f_c > 300$ MHz, $a(h_r) = 3.2(\log_{10}(11.75h_r))^2 - 4.97$ dB. $a(h_r)$ for other operation scenario is also available. Note that the unit for f_c is in MHz and for d is in km, as the model was targeted at outdoor environment. Following a similar approach, empirical models for other frequency range and indoor environment are also established.

While ray-tracing and empirical modeling approaches may lead to more accurate prediction of path loss, the resulting models are quite complex and not suitable for high-level system analysis. Noting that the predicted path loss value in dB scale for both free space and empirical models can be written as a linear function of the logarithm of the distance d, the *log-distance path loss model* is developed and widely used for high-level system analysis. Specifically, path loss at a distance d in dB scale is predicted using the following formula

$$PL(d)\text{dB} = PL(d_0)\text{dB} + 10\alpha \log_{10}(d/d_0), \tag{2.10}$$

where d_0 is the reference distance, $PL(d_0)$ dB is the path loss at d_0, and α is the path loss exponent. The reference distance d_0 is typically selected according to the application environment. For example, d_0 is usually set to 1–10 m for indoor applications and 1 km for outdoor applications. The path loss at d_0 may be determined using the free space path loss model. Finally, the path loss exponent can be estimated by minimizing the mean square error (MSE) between measurement data and the model. Typical values of path loss exponents for different environment are listed in Table 2.1. The following example illustrates the process of establishing a log-distance path loss model.

Example: Log-distance path loss model
Consider an indoor wireless system operating over 900 MHz. The following path loss measurement has been collected for the propagation environment. Establish a log-distance path loss model for the environment.

Table 2.1   *Typical path loss exponent values for different operating environment*

Environment	Typical α values
Urban	4.5
Suburban	4
Rural	3–3.5
Home	3

Distance from transmitter d_i (m)	Path loss values $PL(d_i)_{\text{meas}}$ (dB)
5	43
10	55
20	63
50	72
80	85

Solutions: We need to determine the values of the model parameters: the reference distance d_0, the path loss at d_0, and the path loss exponent α. The reference distance d_0 is chosen to be 2 m considering the indoor environment. The path loss at d_0 is determined using the free space propagation model as

$$PL(d_0) \text{ dB} = 10\log_{10}((4\pi)^2) - 20\log_{10}(1/3) + 20\log_{10}(2) = 37.5 \text{ dB.} \quad (2.11)$$

To calculate the path loss exponent α, we first determine the MSE between the path loss model, now given by $PL(d) \text{ dB} = 37.5 + 10\alpha \log_{10}(d/2)$, and the measurement data as

$$\text{MSE}(\alpha) = \frac{1}{5}\sum_{i=1}^{5}(37.5 + 10\alpha \log_{10}(d/2) - PL(d_i)_{\text{meas}})^2. \quad (2.12)$$

After proper substitution and simplification, the MSE can be shown to be a quadratic function of α given by

$$\begin{aligned}
\text{MSE}(\alpha) = \frac{1}{5}[&(37.5 + 3.98\alpha - 43)^2 + (37.5 + 6.99\alpha - 55)^2 \\
&+ (37.5 + 10\alpha - 63)^2 + (37.5 + 13.98\alpha - 72)^2 \\
&+ (37.5 + 16.02\alpha - 85)^2] \\
= 123.36&\alpha^2 - 656.99\alpha + 886.65 \quad (2.13)
\end{aligned}$$

Table 2.2 Typical values of attenuation factors due to partitions

Building materials	Factor values (FAF/PAF) (dB)
Cloth	1.4
Plasterboard	3.4
Concrete	13
Metal	26

The optimal value for α can be determined by taking derivative with respect to α and setting the result to zero, which leads to

$$\frac{d\text{MSE}(\alpha)}{d\alpha} = 2 \cdot 123.36\alpha - 656.99 = 0. \tag{2.14}$$

α is then determined to be 2.66. Finally, the log-distance path loss model is established as

$$PL(d) \text{ dB} = 37.5 + 26.6 \log_{10}(d/2), \tag{2.15}$$

which can be used to predict path loss at an arbitrary distance d.

Most path loss models apply directly to outdoor propagation environment. For indoor environment, certain correcting factors for partition and flooring loss should be added. These correcting factors vary with the building materials. For example, a cloth partition typically leads to 1.4-dB additional loss, whereas concrete wall may create as much as 13-dB loss. The largest partition loss results from metal partitions, which can be as high as 26 dB. In general, building penetration loss for windows is 6 dB less than that for walls. With the correcting factor, the path loss can be predicted as

$$PL(d)\text{dB} = PL(d_0)\text{dB} + 10\alpha \log_{10}(d/d_0) + \sum_{i=1}^{N_f} \text{FAF}_i + \sum_{i=1}^{N_p} \text{PAF}_i, \tag{2.16}$$

where FAF_i is the ith floor attenuation factor, N_f is the number of floors to traverse, PAF_i is the ith partition attenuation factor, and N_p is the number of partition to traverse. Typical value of attenuation factors for different building materials are listed in Table 2.2.

2.3 Interference analysis for cellular systems

In this section, we apply the path loss models to the cochannel interference analysis for cellular wireless systems. The performance of wireless communication systems are affected by interference generated by transmissions over the same frequency range.

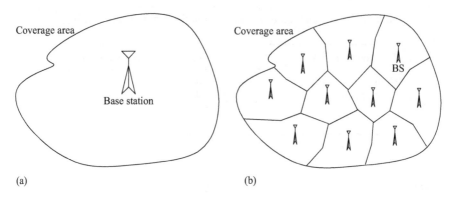

Figure 2.4   The cellular concept: (a) pre-cellular systems and (b) cellular systems

Transmission over unlicensed spectrum can occur randomly in time and/or in space, which acts as interference to each other. Interference also exists for wireless systems using licensed spectrum, as the result of *frequency reuse*. Reuse the same spectrum over geographically separated areas will help greatly improve the capacity of wireless systems. The cellular wireless systems were built exactly based on this idea.

2.3.1   Cellular concept

Conventional wireless systems, including AM/FM radio broadcasting and TV broad-casting systems, use a single-base station to serve all users in the target coverage area. To ensure sufficient coverage of the target area, the base station typically use high transmission power and large antenna, as illustrated in Figure 2.4(a). Should we adopt the same implementation strategy for mobile telephone services, then the single base station will use all available channels to serve eligible mobile users. Such structure is still used in certain specialized systems, such as police or taxi dispatching systems. When more system capacity is needed to support growing demand, the system oper-ators have limited options. The operators can either acquire more radio spectrum or shrink channel bandwidth, to create additional channels. Neither option is scalable as the radio spectrum suitable for wireless transmission is becoming increasingly scarce and the channel bandwidth cannot be made arbitrarily small.

As an illustrative example, let us consider a wireless communication system with total available spectrum of 15 MHz. The channel bandwidth for one-way voice channel is assumed to be 50 kHz. The total number of voice channels that the spectrum can support is 300. With conventional pre-cellular implementation, a single base station will manage all the channels. As such, as many as 150 users can simultaneously communicate with the base station in a full duplex fashion. If the system needs to support more simultaneous full duplex voice communication sessions, the system operators may purchase additional spectrum. Another 5 MHz spectrum can support additional 100 one-way voice channels. Considering the increasingly limited available radio spectrum, such solutions become more and more expensive, if not impossible.

Alternatively, the system operator may adopt advanced transmission technologies, which may reduce the channel bandwidth. If, for example, the bandwidth of one-way voice channel is reduced to 30 kHz, then the same 15 MHz will be able to support 250 full duplex channels. Still, there is a lower limit on channel bandwidth for voice traffic.

The cellular concept was first developed in Bell laboratories in the late 1970s. By exploiting the phenomenon of dramatic signal power attenuation with increasing propagation distance, the cellular concept greatly improves the scalability of wireless systems. With cellular implementation, the target coverage area is partitioned into nonoverlapping small areas, usually referred to as *cells*. Each cell is served by a small base station, as illustrated in Figure 2.4(b). The available channels are divided into channel subsets and each cell is allocated with a channel subset. Typically, the number of cells is larger than the number of channel subsets. Different cells may use the same channel subsets. Cells using a common channel subset are referred to as *cochannel cells*. Since the base stations of cochannel cells user the same channel subset to serve their users, their transmission will create interference to each other, resulting the so-called *cochannel interference*. To control the effect of cochannel interference, cochannel cells are typically separated by cells using different channel subsets. Neighboring cells are usually allocated with different channel subsets. The minimum set of cells that collectively use up all available channels is called a *cluster*, whereas the number of cells in a cluster, or equivalently, the minimum number of required channel subsets, usually referred to as *cluster size*, becomes an important design parameter for cellular systems.

Let us assume that the number of available channel is L. Without cellular concept, as many as L users can communicate at the same time. We now apply the cellular concept by dividing the target coverage area into S cells and allocate channel subsets to different cells. If we choose the cluster size to be N, then we will need N channel subsets, each with L/N channels. As a result, L/N users can communicate at the same time in each cell. Then the total number of users that the cellular system can simultaneously support becomes $S \cdot L/N$. Compared with pre-cellular system, the system capacity is increased by a factor of S/N. Furthermore, the system becomes more scalable because one can increase system capacity by (i) increasing the number of cells in the target area S and (ii) reducing the number of channel subsets (or the cluster size) N. Increasing S has been widely used in the industry with mechanisms such as sectoring, cell splitting, and introducing micro/pico/femto cells. On the other hand, the smallest N depends on the maximum tolerable cochannel interference level. We now perform cochannel interference analysis for cellular system to develop the guideline for determining the smallest N.

2.3.2   Cochannel interference analysis

Let us consider the transmission to a mobile user in the target cell as shown in Figure 2.5, where there are M cochannel cells. While both uplink and downlink transmissions will suffer cochannel interference, the downlink transmission will be more vulnerable, as their experienced interference is coming from the base stations

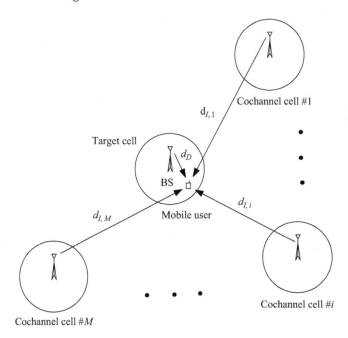

Figure 2.5　Cochannel interference analysis for cellular downlink transmission

of cochannel cells. The severity of cochannel interference can be measured by signal-to-interference ratio (SIR). The SIR at the mobile user can be written as

$$\text{SIR} = \frac{P_0}{\sum_{i=1}^{M} P_{I,i}}, \qquad (2.17)$$

where P_0 is the received signal power and $P_{I,i}$ is the received interference power from ith cochannel cell. The median received signal power at a certain distance from the transmitter can be predicted using the path loss model. Specifically, based on the log-distance path loss model presented in previous section, the path loss experienced by receiver at distance d in linear scale, $PL(d)$, is given by

$$PL(d) = Kd^{\alpha}, \qquad (2.18)$$

where the constant K is equal to $PL(d_0)/d_0^{\alpha}$. Therefore, $PL(d)$ is proportional to d^{α}, where α is the path loss exponent of the environment. As such, the received signal/interference power at distance d is proportional to $d^{-\alpha}$. It follows that the SIR at the mobile user can be calculated as

$$\text{SIR} = \frac{d_D^{-\alpha}}{\sum_{i=1}^{M} d_{I,i}^{-\alpha}}, \qquad (2.19)$$

where d_D is the distance to the serving base station and $d_{I,i}$ is the distance to the ith cochannel base station. Therefore, cellular system should be designed such that the

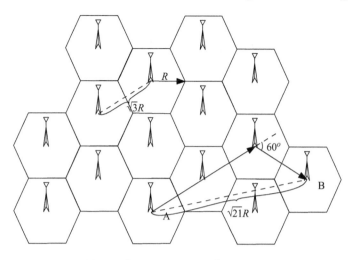

Figure 2.6   Hexagon cell structure

experienced SIR for any mobile users in the coverage area is greater than a minimum required value, denoted by SIR_0. In the following, we demonstrate the selection of the smallest feasible cluster size N to satisfy such design requirements. For that purpose, we first present some results on cell geometry, which facilitate channel subset allocation and SIR calculation.

Cell geometry has great impact on cellular system design and analysis. Note that among the regular shapes, only triangle, square, and hexagon can tessellate a surface, i.e., cover without overlap and gap, whereas hexagon best approximates a circle. As such, each cell with omnidirectional antenna at base station is typically approximated by a regular hexagon. Figure 2.6 shows a typical cell layout with hexagon cells. In particular, each cell is modeled by a hexagon with radius R. Then the center-to-center distance between neighboring cells is $\sqrt{3}R$. It can be further shown that the center-to-center distance between arbitrary two cells is given by

$$D = \sqrt{3}R\sqrt{i^2 + j^2 + ij}, \tag{2.20}$$

where i and j are nonnegative integers, related to how to travel from one cell to the other. Note that one can reach to another cell from current cell by first traveling in one direction for i cells, then turning $60°$ and traveling another j cells. This process is illustrated in Figure 2.6. The distance between cells A and B is equal to $D = \sqrt{3}R\sqrt{2^2 + 1^2 + 1 \cdot 2} = \sqrt{21}R$.

On the basis of these geometric results of hexagon cells, we can ensure that the cochannel cells are of the same minimum distance from each other by always allocating the same channel subsets to cells of distance D apart. In particular, we start from an arbitrary cell and allocate a channel subset. Then we can find all the cells that are of distance D from the initial cell and allocate the same channel subset. Note that each cell will have six distance-D neighbors based on hexagon cell structure, which

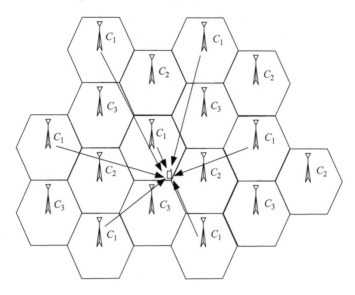

Figure 2.7   Channel subsets allocation for hexagon cell structure

forms a super hexagon. After finding all cochannel cells of the initial cell, we can pick another cell and allocate a different channel subset. The process can be repeated if there are remaining cells without channel subset allocation. A sample channel subset allocation result for hexagon cell structure is shown in Figure 2.7. Cochannel cells are of distance $D = 3R$ away from each other with $i = j = 1$. The number of channel subsets needed is 3. It can be shown that the required number of channel subsets (cluster size) N will be equal to $i^2 + j^2 + ij$, where i and j are the same nonnegative integers used to calculate D.

We now continue our SIR analysis based on the hexagon cell structure. For downlink transmission, the worst case SIR occurs when the mobile user is at the edge of its serving cell. Since the desired signal will experience the largest path loss, the resulting SIR will be the smallest. The interference experienced by the mobile user will be dominated by the transmission from six closest cochannel base stations. As such, the SIR of the downlink transmission is approximately given by

$$\text{SIR} = \frac{R^{-\alpha}}{\sum_{i=1}^{6} d_{I,i}^{-\alpha}}, \tag{2.21}$$

where $d_{I,i}$'s are the distances from the six interference base stations to the mobile user. When D is much greater than R, $d_{I,i}$ can be commonly approximated as D,[1] which lead to the following approximate SIR expression

$$\text{SIR} = \frac{R^{-\alpha}}{\sum_{i=1}^{6} D^{-\alpha}} = \frac{1}{6}\sqrt{3N}^{\alpha}. \tag{2.22}$$

---

[1]The distances $D + R$, $D + R$, D, D, $D - R$, and $D - R$ would provide better approximation, especially when D is not much larger than R, but results in more complex SIR expression.

Table 2.3   *Possible value of cluster sizes*
for hexagon cell structure

i	1	1	2	2	3	3	$\cdots$
j	0	1	0	1	0	1	$\cdots$
Cluster size N	1	3	4	7	9	13	$\cdots$

Therefore, to meet the target SIR requirement, SIR_0, the cluster size N need to be at least $\frac{1}{3}(6 \cdot \text{SIR}_0)^{2/\alpha}$. Note also that the cluster size N should also be equal to $i^2 + j^2 + ij$. The possible values of N is listed in Table 2.3. The following example illustrate the application of these relationship in determining the smallest feasible cluster size N.

Example: Cochannel interference analysis

A wireless communication service provider has enough spectrum to support 1,000 simultaneous two-way voice communication sessions. With pre-cellular implementation, as many as 1,000 mobile users can communicate with the base station simultaneously. The service provider decided to implement a cellular system with frequency reuse. The target area is divided into 420 cells, which are approximated by regular hexagons. The minimum required SIR in the worst case scenario is 10 dB. (i) Assuming a path loss exponent of 2 for the environment, what is the smallest feasible cluster size we can use? (ii) How many mobile users in the coverage area can communicate simultaneously? (iii) What if the path loss exponent is 3?

Solutions:

(i)   To satisfy the worst case minimum SIR requirement, the cluster size N needs to satisfy

$$N \geq \frac{1}{3}(6 \cdot \text{SIR}_0)^{2/\alpha} = \frac{1}{3}(6 \cdot 10)^{2/2} = 20. \tag{2.23}$$

Note that N should equal to $i^2 + j^2 + ij$ for integer i and j. N cannot be 20. Therefore, we choose cluster size $N = 21$, which corresponds to the case of $i = 4$ and $j = 1$.

(ii)   The total number of channel subsets needed is also equal to 21, with $1{,}000/21 \approx 47$ channels in each subset. Therefore, 47 mobile users can communicate at the same time in each cell. The total number of users in the system can communicate simultaneously is $47 \times 420 = 19{,}740$.

(iii)   If $\alpha = 3$, then $N \geq \frac{1}{3}(6 \cdot 10)^{2/3} = 5.1$. We choose cluster size 7, which leads to $1{,}000/7 \approx 143$ channels per cell. The total number of users that the system can support becomes $143 \times 420 = 60{,}060$.

Larger path loss exponent, i.e., faster power attenuation over distance, allows for more frequency reuse and larger system capacity. This observation is widely explored in the capacity enhancement of cellular systems in urban area.

2.4   Shadowing model

Shadowing effect characterizes the blockage of large objects, such as buildings, trees, trucks, and even human body, in the propagation environment. Without shadowing effect, the mobiles of the same distance from the base station will experience identical path loss. As such, the constant received power contour will be an ideal circle. Due to the existence of large objects in the environment, the received power at the mobiles of the same distance from the base station may be different. Considering both path loss and shadowing effect together, the constant received power contour become random irregular shape, as illustrated in Figure 2.8. As the size, position and properties of the objects that affect the target mobile are in general unknown, shadowing effect is typically described in a statistical sense.

The most popular shadowing model is the log-normal model, which has been empirically confirmed. Under the log-normal shadowing model, the path loss in dB scale at distance d, denoted by ψ_{dB}, is modeled as a Gaussian random variable with probability density function (PDF) given by

$$p_{\psi_{dB}}(x) = \frac{1}{\sqrt{2\pi}\sigma_{dB}} \exp\left(-\frac{(x - \mu_{dB})^2}{2\sigma_{dB}^2}\right), \tag{2.24}$$

where μ_{dB} is the average path loss at distance d, predicted with appropriate path loss models from earlier section, and σ_{dB} is the shadowing standard deviation, which varies from 3 to 13 dB depending upon the operating environment. Log-normal model is so

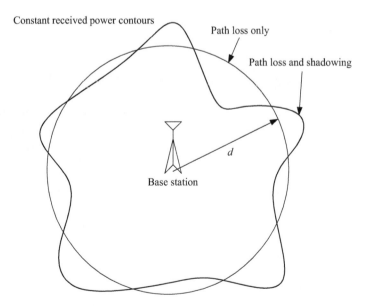

Figure 2.8   Shadowing effect on the constant received signal power contours

named as the path loss in linear scale ψ is a log-normal random variable, with PDF given by

$$p_\psi(x) = \frac{10/\ln(10)}{\sqrt{2\pi}x\sigma_{dB}} \exp\left(-\frac{(10\log_{10}x - \mu_{dB})^2}{2\sigma_{dB}^2}\right). \tag{2.25}$$

The shadowing standard deviation σ_{dB} can be estimated as the square-root of the minimum MSE (MMSE) between the path loss model and the measurement.

Recall from the example on log-distance path loss model, we determine the value of path loss exponent α by minimizing the MSE between measurements and path loss model, given by (2.12). The resulting MMSE, equal to 11.9 after plugging in $\alpha = 2.66$, serves as a quantification of the deviation of the measurement from the path loss model. If we credit such deviation to the shadowing effect, then the MMSE can be used to estimate the shadowing standard deviation σ_{dB} for the log-normal shadowing model. For the specific example under consideration, σ_{dB} is estimated to be $\sqrt{11.9} = 3.45$ dB.

When the log-distance path loss model is applied to predict the average path loss at a certain distance, we arrive at a combined path loss and shadowing model. The shadowed path loss at a distance d from the transmitter is modeled by a Gaussian random variable with mean $PL(d)\text{dB} = PL(d_0)\text{dB} + 10\alpha\log_{10}(d/d_0)$ and variance σ_{dB}^2. The PDF of the shadowed path loss is given by

$$p_{\psi_{dB}}(x) = \frac{1}{\sqrt{2\pi}\,\sigma_{dB}} \exp\left(-\frac{(x - PL(d_0)\text{dB} - 10\alpha\log_{10}(d/d_0))^2}{2\sigma_{dB}^2}\right). \tag{2.26}$$

The combined path loss and shadowing model can be applied to solve important wireless system design problem, as illustrated in the next section.

2.5 Outage and coverage analysis

An important design objective of wireless communication systems is to ensure the system can provide sufficient coverage of the target area. When the received signal power at a particular area is below a certain minimum required power level, the mobile receiver cannot maintain connection with the base station, and therefore, enters an outage status. The probability of entering the outage status, usually referred to as *outage probability*, is an important performance measure for wireless systems.

The outage probability is defined as the probability that the received signal power at the receiver, P_r, falls below a minimum required power level P_{min}. Mathematically, the outage probability, denoted by P_{out}, is defined as

$$P_{out} = \Pr[P_r < P_{min}]. \tag{2.27}$$

With the consideration of the shadowing effect, the received signal power at a particular distance from the transmitter P_r is calculated as $P_t - \psi_{dB}$, where ψ_{dB} is the shadowed path loss. As a result, the outage probability can be calculated as

$$P_{out} = \Pr[P_t - \psi_{dB} < P_{min}] = \Pr[\psi_{dB} > P_t - P_{min}]. \tag{2.28}$$

Using the PDF of ψ_{dB} under the combined path loss and shadowing model developed in previous section, the outage probability can be calculated as

$$
\begin{aligned}
P_{\text{out}} &= \int_{P_t - P_{\min}}^{\infty} p_{\psi_{dB}}(x) dx \\
&= Q\left(\frac{P_t - P_{\min} - PL(d_0)\text{dB} - 10\alpha \log_{10}(d/d_0)}{\sigma_{dB}}\right),
\end{aligned}
\tag{2.29}
$$

where $Q(\,\cdot\,)$ denote the Gaussian Q-function defined as

$$
Q(x) = \int_x^{\infty} \frac{1}{\sqrt{2\pi}} \exp\left(-\frac{z^2}{2}\right) dz.
\tag{2.30}
$$

Further details about Q-function can be found in Appendix A.2. Therefore, the outage probability depends on the transmission power P_t, the receiver sensitivity $P_{\min}$, and the propagation distance d, in addition to the path loss and shadowing effects.

Example: Outage probability

Consider an indoor wireless system deployed in a propagation environment with path loss and shadowing effects characterized by a combined path loss and shadowing model with following parameters: reference distance $d_0 = 2$ m; path loss at d_0 without shadowing effect $PL(d_0)$ dB $= 37.5$ dB; path loss exponent $\alpha = 2.66$; and shadowing standard deviation $\sigma_{dB} = 3.45$ dB. Assume the transmit power of the base station P_t is 10 mW and the minimum required power level at the receiver $P_{\min}$ is -90 dBm. Determine the outage probability for mobile receivers that are of distance 280 m from the base station.

Solutions: With the given model parameters, the average path loss at distance 280 m can be determined as $37.5 + 26.6 \log_{10}(280/2) = 94.6$ dB. Without shadowing effect, the received signal power at the mobile will be 10 dBm $- 94.6$ dB $= -84.6$ dBm, which is greater than $P_{\min}$. The outage probability due to shadowing can be calculated as

$$
P_{\text{out}} = Q\left(\frac{10 - (-90) - 94.6}{3.45}\right) = Q(1.56).
\tag{2.31}
$$

After looking up the Q-function Table A.2.1, we can determine the outage probability to be 0.06.

When a wireless communication system is being deployed to provide mobile communication services, we need to ensure with high probability that the users within the target coverage area enjoy acceptable received signal power level. Such problem is often referred to as cell coverage planning. With combined path loss and shadowing model, we can address such system design problem and arrived at important design guideline. For example, we can answer questions such as for a given transmitting power and target service area, what is the percentage of coverage after considering the path loss and shadowing effect? or for a given target area and a desired coverage percentage, what is the minimum transmission power we can use?

The cell coverage analysis can be related to earlier outage probability analysis. Specifically, if we claim that a location of distance r from the serving base station is covered when the received signal power after shadowing is above the threshold P_{min}, then the coverage probability of this location can be simply calculated as $1 - P_{out}(P_{min}, r)$, where $P_{out}(P_{min}, r)$ is the outage probability. Then, the percentage of coverage can be calculated by averaging the coverage probability at an incremental area over the whole cell. Mathematically, the percentage of coverage, denoted by κ, is calculated as

$$\kappa = \frac{1}{\pi R^2} \int_0^{2\pi} \int_0^R \Pr[P_r(r) > P_{min}] r\,dr\,d\theta$$

$$= \frac{1}{\pi R^2} \int_0^{2\pi} \int_0^R (1 - P_{out}(P_{min}, r)) r\,dr\,d\theta \tag{2.32}$$

where R is the radius of the target coverage area, assumed to be a perfect circle, and $P_r(r)$ is the received signal power at a distance r from the serving base station. With the combined log-distance path loss model and log-normal shadowing model, after substituting the outage probability expression given in (2.29) and much manipulation, the coverage percentage κ can be calculated as

$$\kappa = Q(a) + \exp\left(\frac{2 - 2ab}{b^2}\right) Q\left(\frac{2 - ab}{b}\right) \tag{2.33}$$

where

$$a = \frac{P_{min} - P_t + PL(d_0)\mathrm{dB} + 10\alpha \log_{10}(R/d_0)}{\sigma_{dB}} \quad \text{and} \quad b = \frac{10\alpha \log_{10}(e)}{\sigma_{dB}}. \tag{2.34}$$

Example: Cell coverage

Continuing from previous example, let us assume the same combined path loss and shadowing model for the propagation environment, i.e., reference distance $d_0 = 2$ m; path loss at d_0 without shadowing effect $PL(d_0)$ dB $= 37.5$ dB; path loss exponent $\alpha = 2.66$; and shadowing standard deviation $\sigma_{dB} = 3.45$ dB. The same base station transmission power $P_t = 10$ dBm and minimum received power level $P_{min} = -90$ dBm. Determine the coverage percentage if the target area is a perfect circle centered at the base station and with radius $R = 300$ m.

Solution: With the given values of system and model parameters, we evaluate a and b as

$$a = (-90 - 10 + 37.5 + 26.6 \log_{10} 150)/3.45 = -1.34,$$

$$b = 26.6 \log_{10} e/3.45 = 3.35. \tag{2.35}$$

Substituting these values into the coverage percentage formula in (2.33) leads to

$$\kappa = Q(-1.34) + \exp\left(\frac{2 - (2(-1.34)3.35)}{3.35^2}\right) Q\left(\frac{2 - (-1.34)3.35}{3.35}\right)$$

$$= Q(-1.34) + \exp(0.98)Q(1.94). \tag{2.36}$$

Noting that $Q(-x) = 1 - Q(x)$, the coverage percentage is calculated after Q-function table look up as $\kappa = 0.92 + 2.66 \times 0.025 = 0.9865 = 98.65\%$.

2.6 Further readings

Further discussion about the propagation mechanisms can be found in [1, Chapter 4]. References [1,2] provide detailed discussion on ray-tracing models and empirical models for the path loss effect. More in-depth analysis on cochannel interference is presented in [3, Chapter 3]. Reference [1, Chapter 3] dedicates to the cellular concept, explaining other issues associated with cellular systems, including handoff strategies, cell splitting, and trunking efficiency.

Problems

1. Consider an indoor wireless LAN with carrier frequency $f_c = 2.4$ GHz and an omnidirectional antenna at the access point. The transmit power of the access point is 15 dBm and the minimum received signal power is -82 dBm. Under the free-space path loss model, determine the coverage radius? What if the carrier frequency becomes 5 GHz?

2. We want to establish a log-distance path loss model for an indoor environment. The carrier frequency of the system is 2 GHz. The following path-loss measurements are available.

Distance d in meters	5	10	25	40	60
Path loss measurements in dB	60	73	80	92	104

   (i) Given the measurements, it is preferable to use reference distance of 1 m. What is the path loss at the reference distance based on the free space model?

   (ii) Determine the path loss exponent α of the environment by minimizing the MSE between log-distance path loss model and the measurements.

   (iii) Predict the path loss at distance 50 m with the established model.

3. The received signal power at distance d from the transmitter for log-distance path loss model can be written as $P_r = P_t \cdot C \cdot d^{-\alpha}$, where α is the path loss exponent

of the environment. Determine the expression of the constant C in terms of d_0, $PL(d_0)$ dB and α.

4. Show that the center-to-center distance between any two cells with hexagon cell structure is given by $D = \sqrt{3}R\sqrt{i^2 + j^2 + ij}$, where R is the radius of the hexagon cell, i and j are nonnegative integers.

5. Measurements have shown that the path loss exponent of an outdoor environment is equal to 3.2. The required minimum SIR for the AMPS systems (first generation cellular system with analog technology) is 18 dB.
   (i) Find the smallest number of channel subsets we can use when deploying AMPS in such environment.
   (ii) If the total number of channels that the available spectrum can support is 1,200, how many channels each cell can have with equal allocation?
   (iii) What if the minimum required SIR is reduced to 12 dB (which is acceptable for second general cellular system with digital technology)?

6. Consider a cellular system with hexagonal cells of radius $R = 1$ km. Suppose the minimum distance between cell centers using the same frequency must be $D = 6$ km to maintain the required SIR.
   (i) Find the required cluster size N.
   (ii) If the total number of channels for the system is 500, find the number of channels that can be assigned to each cell.

7. Show that the number of channel subsets needed with hexagon cell structure, i.e., the number of cells in a cluster is equal to $N = i^2 + j^2 + i \cdot j$.

8. Assume the following combined path loss/shadowing model has been established for an outdoor environment: $d_0 = 100$ m, $PL(d_0)$ dB $= 38$ dB, $\alpha = 3.5$, and $\sigma_{dB} = 2.5$ dB.
   (i) Determine the probability that the path loss at distance 2 km is greater than 80 dB.
   (ii) What is the smallest transmission power that the transmitter can use such that the probability that the received signal power at the receiver 3 km away is less than -110 dBm is at most 0.01?

9. Find the cell coverage percentage for a microcellular system with cell radius of 100 m. The path loss follows the log-distance model with $\alpha = 3$, $d_0 = 1$ m, and $PL(d_0)$ dB $= 0$ dB. The shadowing effect follows log-normal shadowing model with standard deviation $\sigma_{dB} = 4$ dB. Assume a transmit power of 80 mW, and a minimum received power requirement of $P_{min} = -100$ dBm.

Bibliography

[1] T. S. Rappaport, *Wireless Communications: Principle and Practice*, 2nd ed. Upper Saddle River, NJ: Prentice Hall, 2002.

[2] A. Goldsmith, *Wireless Communications*, New York, NY: Cambridge University Press, 2005.

[3] G. L. Stüber, *Principles of Mobile Communications*, 2nd ed. Norwell, MA: Kluwer Academic Publishers, 2000.

Chapter 3

Multipath fading

Fading characterizes the effect of random superposition of signal copies received from different propagation paths. These signal replicas may add together constructively or destructively, which leads to a large variation in received signal strength. Multipath fading manifests itself in a much smaller spatial scale than path loss and shadowing, and therefore, is called the small-scale effect of wireless channel. In this chapter, we first develop the general model for multipath fading channels. Then, we discuss the classification of fading channels based on their time domain and frequency domain characteristics. Finally, we present simplified models for two important types of multipath fading channels, which are widely used in the design and analysis of digital wireless transmission systems.

3.1 General fading channel model

Wireless communication systems transmit and receive real bandpass signals with frequency components centered at carrier frequency f_c. Such real bandpass signals can be generally written as

$$s(t) = s_I(t) \cos 2\pi f_c t - s_Q(t) \sin 2\pi f_c t, \tag{3.1}$$

where $s_I(t)$ is the in-phase component and $s_Q(t)$ the quadrature component. To facilitate the development of general fading channel model, we can write a real bandpass signal as the real part of a complex signal as

$$s(t) = \text{Re}\{u(t)e^{j2\pi f_c t}\}, \tag{3.2}$$

where $u(t) = s_I(t) + js_Q(t)$ is the complex envelop of the bandpass signal.

There usually exist multiple propagation paths between the transmitter and the receiver, as illustrated in Figure 3.1. Let us assume that at the time instant t, there are $N(t)$ propagation paths, whereas the nth path introduces a delay $\tau_n(t)$, an amplitude scaling factor $\alpha_n(t)$, and a phase shift $\varphi_n(t)$ to the transmitted signal $s(t)$. Therefore, the received signal over the nth path is given by

$$r_n(t) = \alpha_n(t)e^{j\varphi_n(t)}s(t - \tau_n(t)). \tag{3.3}$$

Applying the complex envelop expression of $s(t)$, $r_n(t)$ can be rewritten as

$$r_n(t) = \text{Re}\{\alpha_n(t)u(t - \tau_n(t))e^{j2\pi f_c(t - \tau_n(t)) + j\varphi_n(t)}\}. \tag{3.4}$$

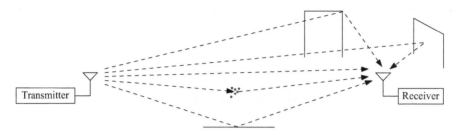

Figure 3.1   Multipath propagation

The received signal copies over different propagation paths will add together noncoherently at the receiver. The overall received signal is given by

$$r(t) = \text{Re}\left\{\sum_{n=1}^{N(t)} \alpha_n(t)u(t - \tau_n(t))e^{j2\pi f_c(t-\tau_n(t))+j\varphi_n(t)}\right\}. \tag{3.5}$$

If we focus on the complex envelops of the transmitted and received signal, we arrive at the following complex baseband input–output relationship for multipath wireless channel, given by

$$\text{Input: } u(t) \Rightarrow \text{Output: } v(t) = \sum_{n=1}^{N(t)} \alpha_n(t)e^{j(-2\pi f_c\tau_n(t)+\varphi_n(t))}u(t - \tau_n(t)). \tag{3.6}$$

Essentially, the complex baseband received signal is the weighted sum of differently delayed version of the complex transmitted signal. The weight of nth copy of the transmitted signal includes an amplitude scaling factor of $\alpha_n(t)$ and a phase shift of $-2\pi f_c\tau_n(t) + \varphi_n(t)$. Note that the phase component $-2\pi f_c\tau_n(t)$ is due to the free space propagation and $\varphi_n(t)$ is the additional phase shift along the nth propagation path. From this input–output relationship, we can determine the impulse response of the complex baseband channel as

$$h(\tau, t) = \sum_{n=1}^{N(t)} \alpha_n(t)e^{j\phi_n(t)}\delta(\tau - \tau_n(t)), \tag{3.7}$$

where $\phi_n(t) = -2\pi f_c\tau_n(t) + \varphi_n(t)$. The above input/output relation implies that the general multipath wireless channel should be modeled as *linear and time-variant (LTV)* systems.

Figure 3.2 illustrates the impulse response of a multipath fading channel. From the illustration, we can observe two major characteristics of multipath fading channels: (i) time-domain variation, as the impulse response changes from time instant t_1 to time instant t_2 and (ii) spread of signal copies along the delay axis, commonly referred to as *delay spread*. The time-domain variation of the channel is caused by the relative motion of the transmitter and the receiver, as well as the changes in the propagation environment. The delay spread is caused by the fact that different propagation paths introduce different amount of delay to the transmitted signal, and as such, the received

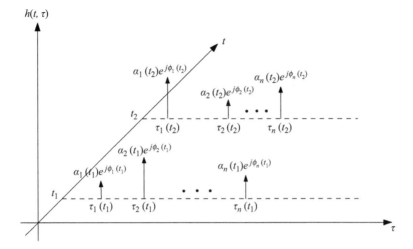

Figure 3.2 Impulse response of time-varying multipath wireless channel

signal becomes the sum of differently delayed copies of the transmitted signal. If the transmitted signal has finite time support, the received signal will have longer time support, due to multipath propagation. In the following sections, we examine these two characteristics of the multipath wireless channels. On the basis of their relative severity with respect to the transmitted signal, the multipath fading channels will be classified to slow/fast fading and frequency flat/selective fading.

Example: Multipath time-varying channel

Consider the outdoor wireless transmission over a multipath channel with carrier frequency 2 GHz. At a particular time instant, there are five propagation paths from the transmitter to the receiver, with path lengths being 1,300, 1,420, 1,450, 1,500, and 1,550 m and corresponding amplitude scaling factors 0.1, 0.02, 0.03, 0.01, and 0.003. Due to propagation effects, such as reflecting, diffraction, and scattering, the transmitted signal experiences additional phase shifts on the second to the fifth path, which are equal to $\pi/2$, $\pi/3$, $2\pi/3$, and $\pi/4$, respectively. Determine the instantaneous impulse response of the channel.

Solutions: From the problem description, we have $N = 5$, $\alpha_1 = 0.1$, $\alpha_2 = 0.02$, $\alpha_3 = 0.03$, $\alpha_4 = 0.01$, $\alpha_5 = 0.003$, $\varphi_1 = 0$, $\varphi_2 = \pi/2$, $\varphi_3 = \pi/3$, $\varphi_4 = 2\pi/3$, and $\varphi_5 = \pi/4$, for the time instant of interest. We need to determine τ_n and the corresponding ϕ_n for each path.

The path delay can be calculated by dividing path length $L_n(t)$ by the propagation speed, which is the speed of light, as

$$\tau_n(t) = L_n(t)/c, \quad n = 1, 2, \ldots, 5. \tag{3.8}$$

As such, we have $\tau_1 = 4.33$, $\tau_2 = 4.73$, $\tau_3 = 4.83$, $\tau_4 = 5$, and $\tau_5 = 5.17$ ns. The corresponding phase shifts are calculated as

$$\phi_n(t) = -2\pi f_c \tau_n(t) + \varphi_n(t), \quad n = 1, 2, \ldots, 5, \tag{3.9}$$

which leads to $\phi_1 = -4\pi/3$, $\phi_2 = -4\pi/3 + \pi/2 = -5\pi/6$, $\phi_3 = -4\pi/3 + \pi/3 = -\pi$, $\phi_4 = 2\pi/3$, and $\phi_5 = -2\pi/3 + \pi/4 = -5\pi/12$. Finally, the instantaneous impulse response of the multipath channel is obtained as

$$\begin{aligned}
h(\tau, t) = {} & 0.1e^{-j4\pi/3}\delta(\tau - 4.33) + 0.02e^{-j5\pi/6}\delta(\tau - 4.73) \\
& + 0.03e^{-j\pi}\delta(\tau - 4.83) + 0.01e^{j2\pi/3}\delta(\tau - 5) \\
& + 0.003e^{-j5\pi/12}\delta(\tau - 5.17).
\end{aligned} \tag{3.10}$$

3.2   Time-domain variation

In this section, we study the rate of time-domain channel variation. Considering the impulse response of the general fading multipath channel given in the previous section, we note that over a short period of time, say in the order of milliseconds, the number of propagation paths $N(t)$ and the amplitude scaling factor of each propagation path $\alpha_n(t)$ will typically remain constant. On the other hand, the phase shift $\phi_n(t) = -2\pi f_c \tau_n(t) + \theta_n(t)$ can vary dramatically as the carrier frequency f_c is normally large. For example, when the carrier frequency f_c is 900 MHz, the phase shift can change by as much as 1.8π if the delay $\tau_n(t)$ changes by 10^{-9} s. Therefore, the speed of time-domain channel variation is governed by the rate of phase shift variation, which is typically measured using *Doppler frequency shift*.

3.2.1   Doppler effect

Doppler effect occurs when there is relative motion between the transmitter and the receiver. Let us focus on received signal at a mobile receiver along a particular propagation path, as illustrated in Figure 3.3. The carrier wavelength of the transmission is λ, which is related to the carrier frequency as $\lambda = c/f_c$, where c is the speed of light. The receiver is moving at the speed of v. The direction of movement and the direction of the incoming radio wave signal form an angle of θ_i. To quantitatively study the rate of phase variation, we calculate the amount of phase change experienced by the transmitted signal after an incremental time period of Δt. Note that after Δt time period, the receiver moves a distance of $v\Delta t$. Assuming the transmitter is far away, the change in the propagation path length after Δt due to the movement is approximately calculated as $\Delta l = v\Delta t \cos(\theta_i)$. It follows that amount of phase change is given by

$$\Delta\phi = 2\pi \frac{v \cos(\theta_i)}{\lambda} \Delta t. \tag{3.11}$$

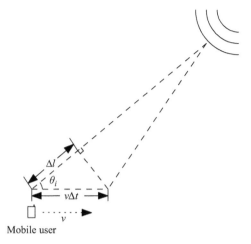

Figure 3.3   Doppler effect due to relative movement

Therefore, the rate of phase variation can be characterized by the Doppler frequency shift, f_{D_i}, given by

$$f_{D_i} = \frac{1}{2\pi} \lim_{\Delta t \to 0} \frac{\Delta \phi}{\Delta t} = \frac{v \cos(\theta_i)}{\lambda}. \tag{3.12}$$

As the result of such frequency shift due to relative motion, the instantaneous frequency of the received signal over the propagation path under consideration becomes $f_c + f_{D_i}$. In general, the transmitted signal will arrive at the receiver along multiple paths from different directions. Obviously, the maximum of Doppler frequency shift occurs when $\theta_i = 0$, which implies the moving direction and the incoming radio wave are opposite to each other. The resulting maximum Doppler shift, i.e., maximum possible spread of transmitted signal bandwidth and also the largest rate of channel phase variation, is given by

$$f_D = v/\lambda. \tag{3.13}$$

In addition, the signal arrived from different paths will experience different amount of Doppler shift. The Doppler effect combined with multipath propagation will increase the signal bandwidth. f_D also characterizes the maximum possible spread of transmitted signal bandwidth due to Doppler effect, as illustrated in the following example.

Example: Doppler effect

Let us assume that a wireless transmitter radiates a sinusoidal signal with frequency 900 MHz. A mobile receiver is traveling at a speed of 75 km/h. As a particular time instant, there are four propagation paths from the transmitter to the receiver. The angles formed by the directions of the incident wave signal and the moving direction θ_i are 15°, 75°, 90°, and 145°, respectively. Determine the frequency components of the received signal at the current time instant.

Solutions: We need to calculate the amount of Doppler frequency shift experienced by signal on different paths. First note that the carrier wavelength λ is equal to $c/f_c = 1/3$ m. The moving speed v is 20.83 m/s. The Doppler frequency shifts are then calculated as

$$f_{D_i} = \frac{20.83 \cos(\theta_i)}{1/3}, \tag{3.14}$$

and given by 60.36, 16.17, 0, and -51.19 Hz. Therefore, the frequency components of the received signal are 900.00006036, 900.00001617, 900, and 899.99994881 MHz. The received signal now effectively has a bandwidth of 111.55 Hz, due to the Doppler effect.

3.2.2 Slow vs fast fading

Maximum Doppler shift f_D essentially characterizes the highest speed of channel phase variation, i.e., the change of $\phi_n(t)$, in multipath propagation environment. Therefore, f_D also serves as a suitable metric to measure the speed of fading channel variation in time domain. Note that larger f_D, typically resulted from large moving speed v and/or smaller carrier wavelength λ, implies faster temporal channel variation. Meanwhile, f_D is a frequency domain metric. In time domain, we use the so-called *channel coherence time* T_c to characterize the speed of time-domain variation. The channel coherence time is defined as the time duration that the channel response remains highly correlated. In terms of the frequency-domain correlation of the channel response, defined as $\rho(f, \Delta t) \triangleq \mathbf{E}[H^*(f,t)H(f,t+\Delta t)]$, where $\mathbf{E}[\,\cdot\,]$ denotes statistical expectation operation and $H(f,t)$ is the time-varying frequency response of the wireless channel and calculated as the Fourier transform of $h(\tau, t)$ with respect to τ, T_c should satisfy $\rho(f, \Delta t) \approx 1$ for all f and $\Delta t < T_c$. It can be shown that T_c is inverse proportional to the maximum Doppler shift $f_D = v/\lambda$, as one would intuitively expect, and can be approximately calculated as $T_c \approx 0.4/f_D$.

With f_D and T_c serve as quantitative measures of channel variation rate in time domain, we can compare them with the properties of the transmitted signal, such as channel bandwidth and symbol period, to determine whether the variation is fast or slow. In particular, we can classify the fading channel into slow fading and fast fading as follows. If the channel coherence time T_c is much greater than the symbol period T_s, then we claim that the transmitted signal experiences slow fading. Otherwise, the fading is considered to be fast. Similarly, if B_s is much greater than f_D, then the Doppler effect is negligible and the transmitted signal experience slow fading.

Example: Slow/fast fading

GSM is a popular second generation cellular system. The channel bandwidth is 200 kHz and the symbol rate on the channel is 270 ksps. Let us consider the outdoor transmission of such system over frequency 2 GHz. Will the transmitted signal to a mobile with moving speed of 75 km/h experience fast or slow fading?

Solutions: Note that the mobile speed is 20.83 m/s and carrier wavelength is $\lambda = c/f_c = 0.15$ m. The maximum Doppler shift f_D is calculated as

$$f_D = v/\lambda = 20.83/0.15 = 138.87 \text{ Hz}. \tag{3.15}$$

The channel coherence time is approximately calculated as $0.4/f_D = 2.88$ ms. The symbol period of GSM transmission is $1/270$ ms, which is much smaller than 2.88 ms. Therefore, the transmitted signal will experience slow fading. In fact, approximately $2.88/(1/270) = 777$ transmitted symbols will experience very similar channel conditions. We can arrive at the same conclusion by comparing B_s with f_D in this example.

To meet the increasing demand for high data rate wireless transmission, most wireless systems will transmit and receive signals with increasingly smaller symbol period T_s and as such operate in a slow fading environment. In such scenario, the channel condition may remain unchanged over a large number of symbol periods. The receiver and even the transmitter may try to acquire certain knowledge of the channel and explore such knowledge for transmission efficiency and performance improvement. Specifically, the transmitter may send some known symbols, usually called *pilot symbols*, periodically to help the receiver estimate the channel response, i.e., $h(\tau, t)/H(f, t)$. The receiver may directly use these channel knowledge to improve the detection performance of subsequently transmitted data symbols. The receiver may also feedback certain channel knowledge to the transmitter to facilitate channel adaptive transmission. These strategies will elaborate further in the following chapters.

In this context, a block fading channel model is often adopted in the design and analysis of various wireless technologies. In particular, the channel response is assumed to remain constant for the duration in the order of the channel coherence time T_c and to become independent afterwards. While serving as an inaccurate approximation of the reality, the block fading channel model greatly facilitates the description and understanding of various wireless transmission technologies, especially those based on channel estimation and feedback. Over a particular coherence time T_c, the complex baseband input–output relationship of multipath wireless channel simplifies to

$$u(t) \Rightarrow v(t) = \sum_{n=1}^{N} \alpha_n e^{-j2\pi f_c \tau_n} u(t - \tau_n). \tag{3.16}$$

The complex baseband impulse response of the channel becomes

$$h(\tau) = \sum_{n=1}^{N} \alpha_n e^{-j2\pi f_c \tau_n} \delta(\tau - \tau_n), \tag{3.17}$$

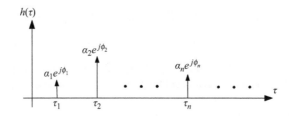

Figure 3.4   Impulse response of linear time-invariant channel

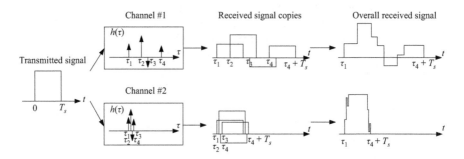

Figure 3.5   Effect of delay spread on transmitted symbol

and is illustrated in Figure 3.4. The fading channel acts as a linear time-invariant filter over each T_c. We now examine the filtering effect of the fading channel as the result of multipath propagation.

3.3   Multipath delay spread

Multipath propagation will introduce time domain dispersion to the transmitted signal. The received signal over different propagation paths will experience different amount of delay, as illustrated in Figure 3.4. The time support of the received signal is larger than that of the transmitted signal by the largest delay difference between propagation paths. Another direct consequence of multipath propagation is the distribution of the received signal power along the delay axis according to the channel response. Each propagation path will introduce a certain amplitude scaling to the transmitted signal. Therefore, the received signal power on different propagation paths will typically be different. The received signal power on the ith path will be proportional to α_i^2, $i = 1, 2, \ldots, N$. Such power spread along the delay axis will have significant effect on the transmitted signals.

Figure 3.5 illustrates the effects of multipath propagation on the transmitted signal over a particular symbol period. Specifically, the same symbol is transmitted over two multipath channels. We can observe that when the power spread along delay axis is significant in comparison to the transmitted symbol duration, as the case with channel #1, the multipath channel introduces severe filtering effect. On the other hand, when

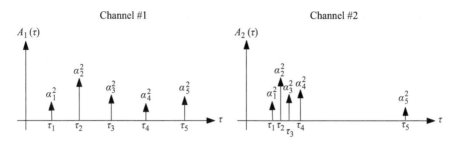

Figure 3.6 Sample power delay profiles

such spread is insignificant, such filtering effect is negligible, as with channel #2. Another observation is that the time support of the received signal will be much larger than the symbol period if the power spread is large compared to T_s, which means that early transmitted symbols will affects the reception of the later transmitted symbols, resulting the so-called intersymbol interference (ISI). Note that such ISI effects also occur in the transmission over bandwidth limited channels. To determine whether the power spread is significant or not, or equivalently, the filtering and ISI effects of the channel is significant or not, we need a proper quantification of the received power spread along the delay axis.

3.3.1 Power delay profile

Power delay profile completely characterizes the distribution of received signal power along delay axis as the result of multipath propagation. Since the received signal power on the ith path is proportional to amplitude gain square, the power delay profile of the channel is usually specified by a plot of power gain α_i^2 vs the delay τ_i for all paths. Figure 3.6 shows two sample instantaneous power delay profiles. The power delay profile of wireless channels remains the same when only the phase shifts of the channel response change. Therefore, power delay profile varies at a much slower rate than the overall channel response. Even when the exact power delay profile changes as the number of multipaths and/or the amplitude of individual path varies, certain properties of the delay spread, such as the average deviation of the spread around the mean delay, will not change dramatically for a given propagation environment. In the following, we present a couple of metrics that are defined based on power delay profile to quantitatively characterize the severity of power spread.

The first metric that we can use to measure the severity of power spread of multipath channels is the so-called maximum delay spread, which is defined as

$$\tau_{\max} = \max_{i,j} |\tau_i - \tau_j|. \tag{3.18}$$

The limitation of this metric is shown in Figure 3.6, where the power delay profile of two channels with five paths from the transmitter to the receiver are shown. Clearly, the delay spread of the first channel is more significant than the second channel. Note that most paths except for one share similar delay for channel #2. Meanwhile, these channels have the same $\tau_{\max} = \tau_5 - \tau_1$. The limitation of $\tau_{\max}$ in measuring delay spread lies in the fact that it fails to take into account the power distribution along delay axis.

The metric RMS (root mean square) delay spread, denoted by σ_T, characterizes delay spread while taking into account the power distribution. σ_T can be interpreted as a measure of the "standard deviation" around the average delay of all paths with the delay of each path being weighted by its power contribution. Specifically, RMS delay spread σ_T can be calculated as

$$\sigma_T = \sqrt{\frac{\sum_{n=1}^{N} \alpha_n^2 (\tau_n - \mu_T)^2}{\sum_{n=1}^{N} \alpha_n^2}}, \tag{3.19}$$

where average delay μ_T is given by

$$\mu_T = \frac{\sum_{n=1}^{N} \alpha_n^2 \tau_n}{\sum_{n=1}^{N} \alpha_n^2}. \tag{3.20}$$

Essentially, the power contribution of different propagation paths are taken into account in the calculation to arrive at a more suitable measure for the severity of delay spread, as illustrated in the following example.

Example: RMS delay spread

The power delay profile of a wireless channel is given in the following tabular format. Determine the maximum delay spread and the RMS delay spread of the channel.

Delay τ (µs)	0.5	1	1.5	2
Power gain α_i^2 (dB)	−20	−10	−30	0

Solutions: The maximum delay spread $\tau_{\max}$ can be easily obtain as $2 - 0.5 = 1.5$ µs. To calculate RMS delay spread σ_T, we first calculate μ_T as

$$\mu = \frac{0.01 \cdot 0.5 + 0.1 \cdot 1 + 0.001 \cdot 1.5 + 1 \cdot 2}{0.01 + 0.1 + 0.001 + 1} = 1.89 \text{ µs}. \tag{3.21}$$

Note that the average delay of all paths without considering their power contribution is equal to 1.25 µs. It follows that we can calculate σ_T as

$$\sigma_T = \sqrt{\frac{\begin{array}{c} 0.01 \cdot (0.5 - 1.89)^2 + 0.1 \cdot (1 - 1.89)^2 \\ + 0.001 \cdot (0.5 - 1.89)^2 + 1 \cdot (2 - 1.89)^2 \end{array}}{0.01 + 0.1 + 0.001 + 1}}$$

$$= 0.315 \text{ µs}. \tag{3.22}$$

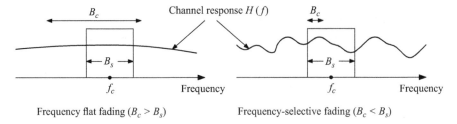

Figure 3.7   Frequency-selective fading vs frequency flat fading

3.3.2   Frequency-flat vs frequency-selective fading

With the RMS delay spread σ_T as the quantification of multipath delay spread for fading channels, we can classify the fading channel into frequency flat and frequency-selective channels. Specifically, if σ_T is large compared to the symbol period T_s, then the delay spread is deemed significant. In this case, the fading channel introduces considerable filtering effect and the transmission will suffer nonnegligible amount of ISI. The channel introduces frequency-selective effect to the transmitted signal, resulting the so-called *frequency-selective fading*. On the other hand, if σ_T is very small compared to T_s, then the delay spread is insignificant, and as such, ISI and channel filtering effect become negligible. In this scenario, we claim that the channel introduces *frequency-flat fading* to the transmitted signal.

This fading channel classification can be better illustrated in the frequency domain, as shown in Figure 3.7. If the channel frequency response varies dramatically over the bandwidth of transmitted signal, B_s, then different frequency components of transmitted signal will be affected differently by the channel. As such, the channel introduces frequency-selective effect. On the other hand, if the channel frequency response remains roughly constant over B_s, then the channel effect on different frequency components of transmitted signal will not differ much. The channel response appears flat to the transmitted signal, i.e., frequency-flat fading.

Multipath channels exhibit more frequency selectivity when the delay spread in time domain is large. As such, RMS delay spread σ_T serves as suitable measure of the frequency selectiveness of multipath channels in time domain. In frequency domain, we use the metric *channel coherence bandwidth*, denoted by B_c, to quantify frequency selectiveness. By definition, channel coherence bandwidth is the bandwidth over which the channel frequency response remains highly correlated. In particular, B_c satisfies: $\rho(\Delta f, t) \stackrel{\Delta}{=} \mathbf{E}[H^*(f, t)H(f + \Delta f, t)] \approx 1$ for all $\Delta f < B_c$. Since significant time-domain delay spread translates to more selectivity in frequency-domain, it can be shown, or intuitively expected, that $B_c \propto 1/\sigma_T$. In fact, B_c can be approximately estimated as $B_c \approx 0.2/\sigma_T$. The wireless channel can then classified as frequency flat if $B_c > B_s$, where B_s is the signal bandwidth. Otherwise, the wireless channel will introduce frequency-selective fading to the transmitted signal.

Example: Frequency flat/selective fading

Let us assume a GSM signal is transmitting over the wireless channel with power delay profile given in previous example. Recall that the channel bandwidth of GSM signal is 200 kHz and the symbol rate on the channel is 270 ksps. Will the transmitted signal experience frequency flat or selective fading?

Solutions: From the previous example, we know that the RMS delay spread of the channel is 0.315 μs. The symbol period of GSM transmission is $1/270 = 3.7$ μs. Since $3.7 \gg 0.315$,[1] the transmitted signal will experience frequency flat fading.

In frequency domain, the channel coherence bandwidth of the channel is calculated as

$$B_c = 0.2/\sigma_T = 0.2/0.315 = 635 \text{ kHz},\qquad\qquad (3.23)$$

which is greater than $B_s = 200$ kHz. We arrive at the same frequency flat fading conclusion.

3.4   Simplified models for fading channels

On the basis of the discussion in previous sections, we can classify fading channels into slow or fast and frequency flat or selective by comparing the characteristics of fading channel and transmitted signal. The classification results are summarized in Figure 3.8. Specifically, the wireless channel can introduce four possible fading effects to the transmitted signal: namely slow frequency flat, slow frequency selective,

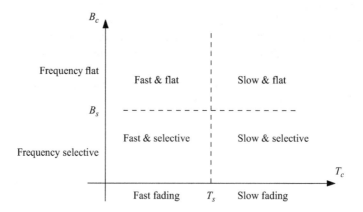

Figure 3.8   Summary of fading classification

[1]Here we adopt the common practice that if $x > 10y$, then $x \gg y$.

fast frequency flat, and fast frequency flat. High data rate wireless systems, e.g., WiFi and cellular systems, have very small symbol periods, and therefore, typically operate in a slow fading environment. Meanwhile, the bandwidth of wireless signal varies dramatically with targeted applications. For example, the channel bandwidth of voice service oriented first generation cellular system is around 30 kHz. The transmitted signal of such system typically experience flat fading. To accommodate higher data rate services, most third generation cellular standards adopt channel bandwidth of several MHz. In forth generation systems, the bandwidth reaches up to 20 MHz. Broadband wireless systems tend to operate in a frequency-selective fading environment.

In this section, we will present the simplified fading channel models for slow frequency flat and slow frequency-selective environment. The slow frequency flat fading case is of great importance as one can effectively convert a frequency-selective fading channel into multiple parallel frequency flat fading channels using multicarrier transmission technology, as will be discussed in the later chapter.

3.4.1 Frequency-flat fading

Recall the complex baseband input–output relationship of multipath wireless channel for slow fading environment, given by

$$u(t) \Rightarrow v(t) = \sum_{n=1}^{N} \alpha_n e^{j\phi_n} u(t - \tau_n), \tag{3.24}$$

where $u(t)$ is the complex baseband transmitted signal and $v(t)$ is the complex baseband received signal. In frequency flat fading environment, the delay spread of multipath channel is insignificant compared to symbol period, i.e., $\sigma_T \ll T_s$. Therefore, different propagation paths introduce approximately the same delay, e.g., $\tau_1 \approx \tau_2 \approx \cdots \approx \tau_N$. Neglecting the delay difference, we arrive at a simplified input–output relationship.

$$u(t) \Rightarrow v(t) = \left(\sum_{n=1}^{N} \alpha_n e^{j\phi_n} \right) u(t - \tau_o), \tag{3.25}$$

where τ_o denotes the common delay of all paths. Noting that the common delay can be compensated by the synchronization process at the receiver, we conclude that for frequency flat fading environment, the effect of fading channel is captured by the complex channel gain

$$z = \sum_{n=1}^{N} \alpha_n e^{j\phi_n}. \tag{3.26}$$

Specifically, the received complex baseband signal over flat fading channel is given by

$$v(t) = z \cdot u(t) + \tilde{n}(t) \tag{3.27}$$

where $\tilde{n}(t)$ is the complex baseband additive noise, usually assumed to be white Gaussian. Therefore, the complex channel gain z is the key factor that differentiate

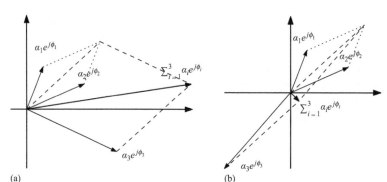

*Figure 3.9   The effect of phase difference on the sum of complex numbers:
(a) constructive addition and (b) destructive addition.*

flat fading channel from conventional additive white Gaussian noise (AWGN) channel, for which the received signal is simply given by

$$v(t) = u(t) + \tilde{n}(t). \tag{3.28}$$

The introduction of the complex channel gain z due to fading brings many new challenges to the wireless system design. In particular, when the power of the transmitted signal is fixed to P_t, the received signal power will be proportional to the amplitude square of the complex channel gain z, as $P_r = |z|^2 P_t$. Therefore, unlike AWGN channels, the received signal power over fading wireless channel is varying over time with the complex gain z. Note that z is the sum of a large number of complex path gains with amplitude α_n and phase ϕ_n, $n = 1, 2, \ldots, N$. The amplitude of z can change dramatically over a short period of time depending on the relative phases of different multipath components. Sometime, different path gains $\alpha_n e^{j\phi_n}$ will add together constructively, leading to relatively large amplitude of z, denoted by $|z|$, as shown in Figure 3.9(a). It may also happen that different path gains $\alpha_n e^{j\phi_n}$ will add together destructively, resulting very small $|z|$, as illustrated in Figure 3.9(b). Due to the Doppler effect, the phase shifts introduced by different paths ϕ_n will vary dramatically over a short period of time, which leads to fast variation of the channel amplitude $|z|$ over time. The temporal variation of $|z|$ is illustrated in Figure 3.10.

The resulting variation of received signal power will immediately affect the received signal-to-noise ratio (SNR) and in turn the detection performance. On the other hand, the complex channel gain z also introduces a phase shift of $\angle z$, which we refer to as channel phase in the following discussion. Such phase shift will be added to the phase of transmitted complex baseband signal $u(t)$. Most modern digital wireless transmission systems use the phase of transmitted signal to carry information. The additional phase shift due to fading wireless channel, if not properly compensated, will result in very poor information detection performance. The effect of $|z|$ and $\angle z$ on digital wireless transmission and their mitigation will be thoroughly discussed in the following chapters.

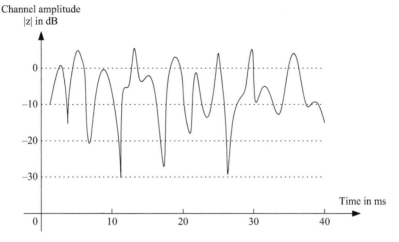

Channel amplitude
|z| in dB

Figure 3.10   Amplitude variation of flat fading channel gain

Example: Flat fading channel

Consider wireless transmission over a multipath channel with carrier frequency 2 GHz. There are five propagation paths from the transmitter to the receiver, with path lengths 1,300, 1,420, 1,450, 1,500, and 1,550 m and corresponding amplitude scaling factors 0.1, 0.02, 0.03, 0.01, and 0.003. Additional phase shifts on the second to the fifth path are $\pi/2$, $\pi/3$, $2\pi/3$, and $\pi/4$, respectively. Assume that the symbol period of transmitted signal is much greater than the delay spread of the multipath channel, and as such, the fading channel introduces frequency flat fading. Determine the instantaneous complex channel gain for the channel. Repeat the calculation assuming the receiver moves to a new location with path lengths changed to 1,300.5, 1,421, 1,452, 1,500.1, and 1,550.2 m and other parameters unchanged.

Solutions: The instantaneous impulse response can obtained (see previous example) as

$$h(\tau, t) = 0.1e^{-j4\pi/3}\delta(\tau - 4.33) + 0.02e^{-j5\pi/6}\delta(\tau - 4.73)$$
$$+ 0.03e^{-j\pi}\delta(\tau - 4.83) + 0.01e^{j2\pi/3}\delta(\tau - 5)$$
$$+ 0.003e^{-j5\pi/12}\delta(\tau - 5.17). \tag{3.29}$$

Since the delay spread is insignificant, we can assume all paths have same delay. The impulse response becomes

$$h(\tau, t) = (0.1e^{-j4\pi/3} + 0.02e^{-j5\pi/6} + 0.03e^{-j\pi}$$
$$+ 0.01e^{j2\pi/3} + 0.003e^{-j5\pi/12})\delta(\tau - \tau_0). \tag{3.30}$$

Therefore, the instantaneous complex channel gain is calculated as

$$z = 0.1e^{-j4\pi/3} + 0.02e^{-j5\pi/6} + 0.03e^{-j\pi} + 0.01e^{j2\pi/3}$$
$$+ 0.003e^{-j5\pi/12} = 0.12e^{j2.57}. \tag{3.31}$$

When the receiver moves to a new location, the lengths of propagation paths change slightly. Such change typically will affect the phase shifts more than the amplitude scaling factor. Following the similar process as in earlier example, we can calculate current phase shifts $\phi_n(t')$ using the relationship

$$\phi_n(t') = -2\pi L_n(t')/\lambda + \varphi_n(t'), \quad n = 1, 2, \ldots, 5, \tag{3.32}$$

where $\lambda = c/f_c = 0.15$ m. The phase shift are determined as $\phi_1(t') = 0$, $\phi_2(t') = -\pi/6$, $\phi_3(t') = \pi/3$, $\phi_4(t') = -2\pi/3$, and $\phi_5(t') = -5\pi/12$. It follows that the instantaneous complex channel gain is calculated as

$$z' = 0.1e^{-j0} + 0.02e^{-j\pi/6} + 0.03e^{-j\pi/3} + 0.01e^{-j2\pi/3}$$
$$+ 0.003e^{-j5\pi/12} = 0.14e^{j0.36}. \tag{3.33}$$

3.4.2 Autocorrelation of complex channel gain

The autocorrelation function of the complex channel gain z characterizes the variation of the flat fading channel over space or time. The complex gain z can be written into the in-phase and quadrature components format as

$$z = z_I + jz_Q = \sum_{n=1}^{N} \alpha_n \cos \phi_n + j \sum_{n=1}^{N} \alpha_n \sin \phi_n. \tag{3.34}$$

When the number of paths N is large, after invoking the central limit theorem and the fact that amplitude α_n and phases ϕ_n for different paths are independent, we can model z_I and z_Q as joint Gaussian random processes. When there is no dominant LOS component, noting that the phase of the nth path $\phi_n = -2\pi f_c \tau_n$ changes rapidly for a small change in τ_n, we can assume that ϕ_n is uniformly distributed over $[-\pi, \pi]$. It follows that the statistical expectation of z_I and z_Q are calculated as

$$E[z_I] = \sum_{n=1}^{N} E[\alpha_n]E[\cos \phi_n] = 0, \qquad E[z_Q] = \sum_{n=1}^{N} E[\alpha_n]E[\sin \phi_n] = 0. \tag{3.35}$$

We can also show that the correlation between z_I and z_Q is given by

$$E[z_I z_Q] = \sum_{n=1}^{N} E[\alpha_n^2]E[\cos \phi_n \sin \phi_n] = 0 \tag{3.36}$$

These results imply that the in-phase and quadrature components of z are independent Gaussian processes with zero mean. When dominant LOS component exists, the uniform distribution assumption on path phases does not hold anymore. z_I and z_Q will have nonzero mean and may be correlated.

The phase shift ϕ_n varies much faster than the number of paths N and the amplitude of each path α_n. We introduce time dependent to the phase shifts and rewrite the in-phase component z_I as

$$z_I(t) = \sum_{n=1}^{N} \alpha_n \cos \phi_n(t). \tag{3.37}$$

The autocorrelation of $z_I(t)$ can be calculated as

$$A_{z_I}(\Delta t) = \mathbf{E}[z_I(t)z_I(t + \Delta t)] = \sum_{n=1}^{N} \mathbf{E}[\alpha_n^2]\mathbf{E}[\cos \phi_n(t) \cos \phi_n(t + \Delta t)]. \tag{3.38}$$

Applying the trio-geometric relationship, we have

$$\mathbf{E}[\cos \phi_n(t) \cos \phi_n(t + \Delta t)]$$
$$= \frac{1}{2}\{\mathbf{E}[\cos (\phi_n(t) + \phi_n(t + \Delta t))] + \mathbf{E}[\cos (\phi_n(t) - \phi_n(t + \Delta t))]\}. \tag{3.39}$$

Since for non-LOS scenario, ϕ_n is uniformly distributed over $[-\pi, \pi]$, $\mathbf{E}[\cos (\phi_n(t) + \phi_n(t + \Delta t))] = 0$. For small enough Δt, the phase difference $\phi_n(t) - \phi_n(t + \Delta t)$ can be shown to be equal to $2\pi f_D \cos \theta_n \Delta t$, where $f_D = v/\lambda$ is the maximum Doppler spread and θ_n is the angle formed by the nth incident path with the relative moving direction between the transmitter and the receiver. The autocorrelation function of $z_I(t)$ simplifies to

$$A_{z_I}(\Delta t) = \frac{1}{2} \sum_{n=1}^{N} \mathbf{E}[\cos 2\pi f_D \cos \theta_n \Delta t]. \tag{3.40}$$

To proceed further, we need to make additional assumption about the propagation environment. Let us consider the uniform scattering environment (also known as Jakes' model), where θ_n is uniformly distributed between 0 and 2π and each path contribute the same amount of power, i.e., $\mathbf{E}[\alpha_n^2] = P_r/N$, where P_r is the total received signal power, we can show that the autocorrelation of z_I becomes

$$A_{z_I}(\Delta t) = \frac{P_r}{2N} \sum_{n=1}^{N} \mathbf{E}[\alpha_n^2]\mathbf{E}[\cos 2\pi f_D \cos \theta_n \Delta t]. \tag{3.41}$$

When N becomes very large and the difference between different θ_ns become very small, $A_{z_I}(\Delta t)$ converges in the limit to

$$A_{z_I}(\Delta t) = \frac{P_r}{2} J_0(2\pi f_D \Delta t), \tag{3.42}$$

where $J_0(x)$ is the first kind Bessel function of order zero, defined as

$$J_0(x) = \frac{1}{\pi} \int_0^\pi e^{-jx \cos \theta} d\theta. \tag{3.43}$$

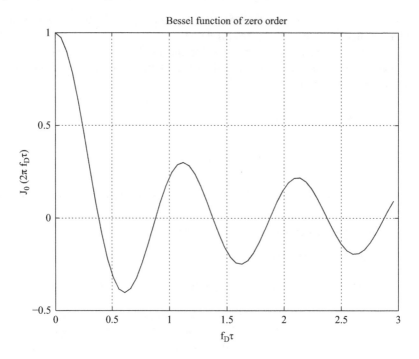

Figure 3.11   Normalized plot of the Bessel function

A normalized plot of $J_0(2\pi f_D \Delta t)$ is shown in Figure 3.11. Similarly, we can show that

$$A_{z_Q}(\Delta t) = \frac{P_r}{2}J_0(2\pi f_D \Delta t), \qquad A_{z_I, z_Q}(\Delta t) = 0. \tag{3.44}$$

It follows that the autocorrelation function of the complex channel gain z is given by

$$A_z(\Delta t) = P_r J_0(2\pi f_D \Delta t), \tag{3.45}$$

which is equal to zero when $f_D \Delta t = 0.38$ and remain relative small when $f_D \Delta t \geq 0.5$, i.e., $v\Delta t \geq 0.5\lambda$. On the basis of this result, we arrive at a rule of thumb that channel gain becomes uncorrelated over a distance of half wavelength. This result also partially justifies the approximate formula used to calculate channel coherence time T_c from the maximum Doppler spread f_D in the previous subsection.

Example: Correlation of complex channel gain

Wireless communication systems often utilize multiple antennas to improve transmission efficiency and reliability. Consider the wireless transmission with carrier frequency 2 GHz to a mobile receiver equipped with three antennas. The antennas are arranged in a linear array with spacing of 30 cm. Determine the correlation between the complex channel gains experienced by signal received on different antennas.

Solutions: The carrier wavelength is $\lambda = c/f_c = 0.15$ m. The correlation coefficient of complex channel gains at two antenna locations can be calculated as

$$\rho(\Delta d) = \frac{A_z(\Delta t)|_{\Delta t = \Delta d/v}}{P_r} = J_0(2\pi(v/\lambda)\Delta d/v) = J_0(2\pi\Delta d/\lambda). \qquad (3.46)$$

Therefore, the correlation of the complex channel gains corresponding to neighboring antennas is $J_0(2\pi 0.2) = 0.64$. The correlation corresponding to nonneighboring antennas is $J_0(2\pi 0.4) = -0.053$.

3.4.3 Frequency-selective fading

High data-rate wireless systems typically have large channel bandwidths and tend to operate in frequency-selective fading environment. In this case, the delay spread of multipath components is significant and the delay difference between paths cannot be neglected. The received complex baseband signal over slow frequency-selective fading channel is given by

$$v(t) = \sum_{n=1}^{N} \alpha_n e^{j\phi_n} u(t - \tau_n) + \tilde{n}(t), \qquad (3.47)$$

where α_n, ϕ_n, and τ_n are the amplitude, phase shift, and delay, respectively, introduced by the nth propagation path and $\tilde{n}(t)$ is the complex baseband noise. The channel essentially acts as a continuous-time linear filter on the transmitted signal. The fact that τ_n is continuous variable makes such model too complex for system analysis and design. Instead, a discrete-time approximation of general selective fading channel model is often adopted.

The discrete-time approximation is based on the fact that the receiver can only resolve multipath with delay difference greater than $1/B$, where B is the channel bandwidth. Specifically, two propagation paths with delays τ_n and τ_m cannot be resolved by the receiver if $|\tau_n - \tau_m| < 1/B$. We divide the delay axis into L equal length intervals of duration $\Delta\tau \approx 1/B$, as illustrated in Figure 3.12. L is typically chosen such that $L\Delta\tau \geq \tau_{\max}$, where $\tau_{\max}$ is the maximum delay spread. Multipath signals arrived at the receiver within the same interval are not resolvable to the receiver. Since their delay difference is relative small, we can assume that these multipath signals arrive at the same time instant $l\Delta\tau$, $l = 0, 1, \ldots, L - 1$, and their composite effect could be characterized by the sum of their individual path gains. Specifically, the composite gain of the paths arriving in the lth interval, denoted by h_l, is given by

$$h_l = \sum_{l\Delta\tau < \tau_i \leq (l+1)\Delta\tau} \alpha_i e^{j\phi_i}, \quad l = 0, 1, \ldots, L - 1. \qquad (3.48)$$

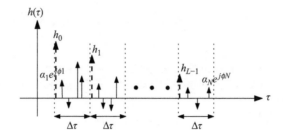

Figure 3.12   *Discrete-time approximation for selective fading*

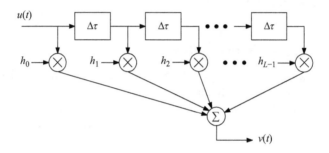

Figure 3.13   *Tapped delay line channel model for selective fading*

With such approximation, the channel impulse response is simplified to

$$h(\tau) = \sum_{l=0}^{L-1} h_l \delta(\tau - l\Delta\tau), \tag{3.49}$$

where L is typically known as the length of the channel. The resulting model for selective fading is often referred to as the *tapped delay line* model, which essentially models the wireless channel as a discrete-time filter. Figure 3.13 illustrates the tapped delay line model for selective fading. h_l is also referred to as the gain of lth tap. The received complex baseband signal over selective fading under this simplified model becomes

$$v(t) = \sum_{l=0}^{L-1} h_l \cdot u(t - l\Delta\tau) + \tilde{n}(t). \tag{3.50}$$

Such frequency-selective fading channel model applies to most broadband wireless transmission system. Meanwhile, most advance wireless transmission technologies, such as diversity combining, channel adaptive transmission, and multi-antenna transmission, are often designed based on the frequency flat fading channel models. The basic premise of such approach is that a wideband frequency selected channel can be converted into multiple parallel frequency flat fading channels using the multicarrier transmission technology. We will present the idea of multicarrier transmission and its discrete implementation, commonly known as orthogonal frequency division multiplexing (OFDM) in Chapter 6.

3.5 Further readings

Further discussion about the correlation function of the channel response can be found in [1, Chapter 3]. Both [2, Chapter 5] and [3, Chapter 3] present several methods to simulate the multipath fading channel in laboratory environment. Reference [4] provides further details of the Jakes propagation model.

Problems

1. Given the in-phase/quadrature representation of a real bandpass signal $s(t) = 2\cos \pi 6{,}000t - \sin \pi 6{,}000t$, determine its equivalent (i) complex baseband representation and (ii) envelop and phase representation.

2. Assume that the bandpass signal $s(t) = 3\cos \pi 6{,}000t - 2\sin \pi 6{,}000t$ was transmitted over a wireless channel, and the received bandpass signal is given by $r(t) = 1.5\cos(\pi 6{,}000t - \pi/6) - \sin(\pi 6{,}000t - \pi/6) - \cos(\pi 6{,}000(t - 2) - \pi/4) + 2/3\sin(\pi 6{,}000(t - 2) - \pi/4)$. Determine the impulse response of the complex baseband equivalent channel.

3. Consider an indoor wireless transmission over 2-GHz frequency. Let us assume that there are five paths between the transmitter and the receiver, with path lengths being 13, 14, 14.5, 15.2, and 15.5 m and corresponding path power gains 0.03, 0.01, 0.005, 0.002, and 0.001. The additional phase shifts for the second to the fifth paths are $\pi/4$, $\pi/5$, $2\pi/5$, and $\pi/6$. Determine the instantaneous impulse response of the channel. Repeat the calculation assuming the mobile moves to a new location, which leads to path lengths changed to 13.1, 14.1, 14.6, 15, and 15.3 m.

4. A mobile is receiving signal from a stationary transmitter over the LOS path. The transmission signal bandwidth is 30 kHz with carrier frequency 950 MHz. Determine the frequency range of the received signal if the traveling speed of the mobile is (i) 1, (ii) 10, and (iii) 100 km/h.

5. A mobile terminal is moving at the speed of 60 km/h while receiving the radio transmission from a base station over carrier frequency 900 MHz. Determine the value range of the symbol period T_s, such that the transmitted signal will experience slow fading.

6. Consider an outdoor wireless transmission over 2-GHz frequency. Let us assume that there are five paths between the transmitter and the receiver, with path lengths being 1,300, 1,420, 1,450, 1,500, and 1,550 m and corresponding path gains 0.01, 0.002, 0.003, 0.001, and 0.0003. Calculate the RMS delay spread of the channel.

7. Compare the maximum delay spread and RMS delay spread of the following channels

Channel #1				
Delay τ_n in µs	0	1	2	3
Power a_n^2 in dB	0	−5	−20	−10

Channel #2				
Delay τ_n in µs	0	0.1	0.5	3
Power a_n^2 in dB	−5	0	−10	−30

     If the symbol rate of the transmitted signal is 1 Msps, will the transmitted signal experience frequency flat or selective fading over each channel?

8.   A wireless transmission system operating over 5.8-GHz frequency range is transmitting with symbol rate 100 ksps. (i) Find the value range of the RMS delay spread of the channel such that the transmitted signal experiences flat fading; (ii) will the channel introduce slow or fast fading if the receiver is moving at a speed of 30 km/h?

9.   Consider an indoor wireless transmission over 2-GHz frequency. Let us assume that there are five paths between the transmitter and the receiver, with path lengths being 13, 14, 14.5, 15.2, and 15.5 m and corresponding path power gains 0.03, 0.01, 0.005, 0.002, and 0.001.

    (i)   Assuming that the symbol period of transmitted signal is much largest than the RMS delay spread, and as such, the transmitted signal experience frequency flat fading. Determine the amplitude and phase of the instantaneous complex channel gain z;

    (ii) Now assume that the receiver moved to a new location, which causes the lengths of first three paths increased by 0.1 m and those of the last two path increased by 0.2 m. What is the amplitude and phase of z now?

10. A mobile is equipped with two antennas that are separated by 8 cm. Determine the minimum value of the carrier frequency that the transmitter should use to ensure that the signals received at two antennas experience approximately uncorrelated fading, i.e., $\rho \le 0.5$.

Bibliography

[1]   A. Goldsmith, *Wireless Communications*, New York, NY: Cambridge University Press, 2005.

[2]   T. S. Rappaport, *Wireless Communications: Principle and Practice*, 2nd ed. Upper Saddle River, NJ: Prentice Hall, 2002.

[3]   G. L. Stüber, *Principles of Mobile Communications*, 2nd ed. Norwell, MA: Kluwer Academic Publishers, 2000.

[4]   W. C. Jakes, *Microwave Mobile Communication*, 2nd ed. Piscataway, NJ: IEEE Press, 1994.

Chapter 4
Digital transmission over flat fading

Most modern wireless communication systems employ digital transmission schemes. With digital transmission, source information are always converted into digital format before being transmission. Besides many other desirable features, digital transmission enjoys following key advantages: (i) facilitate source/channel coding for efficient transmission and error protection; (ii) provide better immunity to additive noise and interference; and (iii) achieve higher spectrum efficiency with guaranteed error performance through adaptive transmission.

This chapter studies the effect of fading channels on digital wireless transmission. After reviewing the basics of digital bandpass modulation, we investigate the effect of channel phase and channel amplitude on the detection performance of linear modulation schemes. Finally, we present the statistical fading channel models for various fading scenarios and apply them to the performance analysis of digital wireless transmission, in terms of outage probability and average error rate.

4.1 Basics of digital bandpass transmission

The structure of a generic digital wireless transmission system is shown in Figure 4.1. The information generated by the source first goes through source encoder, which removes the redundancy of the source information as well as performs analog-to-digital conversion, if necessary. The channel encoder then adds some controlled redundant bits to provide error detection and correction capability. The resulting coded information, in the form of binary bit sequence, is then processed by the modulator. The objective of the modulator is to convert the coded binary information into a form that is suitable for transmission over the channel, which is typically sinusoidal for wireless communications. After that, the modulated signal will go through RF processing, including power amplification, before being transmitted from the antenna.

After propagating through the wireless channel, the transmitted signal reaches the receive antenna, but with much lower power level. The RF circuits at the receiver apply operations, including bandpass filtering, low-noise amplifying, and down-conversion, to the received signal collected by the antenna. Then, the down-converted signal is processed by the demodulator that essentially performs the demapping operation from waveforms to binary data. In particular, during each symbol period, the demodulator will decide the symbol that was most likely transmitted based on the received signal

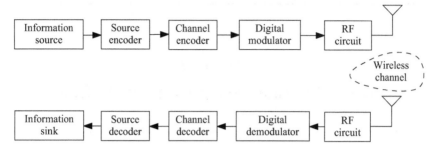

Figure 4.1   Digital wireless transmission system

and determine the corresponding coded bit sequence. The demodulator output will then be processed by channel decoder, which performs error detection and correction if necessary. The source decoder will convert the decoded bit sequence to a format acceptable to the information sink.

The modulation/demodulation scheme is the most essential characteristic of a digital transmission system. Different modulation schemes lead to different trade-offs among spectrum efficiency, power efficiency, error performance, and implementation complexity. The desired properties of a modulation scheme for wireless transmission include: (i) high spectral efficiency to better explore the limited spectrum resource; (ii) high power efficiency to preserve the valuable power resource of the battery powered mobile terminals; (iii) robustness to the impairments introduced by multipath fading; (iv) low implementation complexity to reduce the overall system cost. Usually, these are conflicting requirements. Therefore, the best choice would be the one resulting the most desirable trade-off.

The digital bandpass modulation process can be viewed as the mapping from information bit/bit sequence to sinusoidal waveforms. Typically, n information bits are mapped to a sinusoidal of time duration T_s, as

$$\{d_j\}_{j=1}^n \implies A_i \cos(2\pi f_i t + \theta_i), \quad 0 \le t \le T_s, \tag{4.1}$$

where d_js are coded information bits, A_i, f_i, and θ_i are the amplitude, frequency, and phase of the sinusoidal, respectively. The sinusoidal is usually referred to as modulated symbol, whereas T_s is called symbol period. To uniquely represent 2^n possible bit sequences of length n, we will need 2^n distinct sinusoidal, which may differ in either amplitude, phase, and/or frequency.

When the modulated symbols differ in their amplitude and/or phase, not in frequency, the modulation process can be implemented with linear operations, resulting the so-called *linear modulation schemes*. On the other hand, if the modulated symbols differ in frequency, we have *nonlinear modulation schemes*. Typically, linear modulation schemes, e.g., quadrature-amplitude modulation (QAM), can achieve high spectrum efficiency, whereas nonlinear modulation schemes, e.g., frequency shift keying (FSK), have higher power efficiency and lower receiver complexity. For the best detection performance, linear modulation schemes typically require *coherent detection*, i.e., the receiver needs to recover the reference phase of the transmitter which implies higher receiver complexity. Nonlinear modulation schemes can be noncoherently detected and therefore lead to lower receiver complexity. Nonlinear modulation schemes were popular in earlier generation of wireless systems. As the demand for higher spectrum efficiency prevails, linear modulation schemes have gain wide acceptance in advanced wireless communication systems.

4.2   Linear bandpass modulation

In this book, we focus on linear bandpass modulation schemes, which are widely used in advanced wireless transmission systems. With linear bandpass modulation, the digital information is carried using either the amplitude and/or phase of the sinusoidal. In particular, the modulated symbol over the ith symbol period can be written as

$$s_i(t) = A_i \cos(2\pi f_c t + \theta_i), \quad 0 \le t \le T_s, \tag{4.2}$$

where f_c is the carrier frequency, A_i and θ_i are information carrying amplitude and phase, respectively. If only the amplitude A_i is changing from symbol to symbol, we have amplitude shift keying (ASK) modulation scheme. If only the phase θ_i differs between symbols, we have phase shift keying (PSK) scheme. When both A_i and θ_i may change from symbol to symbol, we have the more general amplitude/phase shift keying (APSK) and QAM schemes.

Applying the trigonometric relationship, the modulated symbols of linear modulation schemes can be rewritten into the in-phase/quadrature representation as

$$s_i(t) = s_I(i) \cos 2\pi f_c t - s_Q(i) \sin 2\pi f_c t, \tag{4.3}$$

where $s_I(i) = A(i) \cos \theta(i)$ is the in-phase component and $s_Q(i) = A(i) \sin \theta(i)$ is the quadrature component. Most linear modulators are implemented based on these in-phase/quadrature representation of the transmitted signal. The generic structure of linear modulator is shown in Figure 4.2. Note that different modulation schemes will differ only in the mapping from bits to in-phase/quadrature components. Such properties greatly facilitate the implementation of adaptive modulation technology, as discussed in the later chapter.

Finally, the modulated symbols $s_i(t)$ can be written into the complex envelop format as

$$s_i(t) = \text{Re}\{A_i e^{j\theta_i} e^{j2\pi f_c t}\}, \quad 0 \le t \le T_s, \tag{4.4}$$

where $A_i e^{j\theta_i}$ is the complex baseband symbol. Essentially, the mapping from bit sequence $\{d_j\}_{j=1}^n$ to sinusoidal for linear modulation schemes can be equivalently characterized by the mapping from bit sequence $\{d_j\}_{j=1}^n$ to the complex baseband symbols $s_i \triangleq A_i e^{j\theta_i}$. Each baseband symbol s_i corresponds to a point in the complex plane defined by $\cos(2\pi f_c t)$ and $\sin(2\pi f_c t)$ with coordinates $(A_i \cos(\theta_i), A_i \sin(\theta_i))$. The collection of all possible symbol points in the complex plane forms a constellation for the modulation scheme. The modulation mapping can also be specified by the mapping from bit sequence $\{d_j\}_{j=1}^n$ to constellation points.

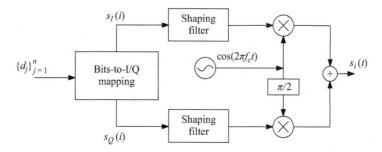

Figure 4.2   Structure of linear modulator based on in-phase/quadrature representation

To illustrate further, let us consider the quadrature PSK (QPSK) modulation scheme, the variants of which have been widely adopted in various wireless standards. With QPSK, each pair of coded bits is mapped to a sinusoidal with common frequency f_c and amplitude A but a unique phase. To ensure the maximal distinction between sinusoidal, the phase difference between neighboring symbols are set to $\pi/2$. Therefore, a possible mapping scheme for QPSK is given by

$$00 \Longrightarrow s_1(t) = A\cos\left(2\pi f_c t + \frac{\pi}{4}\right);$$

$$01 \Longrightarrow s_2(t) = A\cos\left(2\pi f_c t + \frac{3\pi}{4}\right);$$

$$11 \Longrightarrow s_3(t) = A\cos\left(2\pi f_c t + \frac{5\pi}{4}\right);$$

$$10 \Longrightarrow s_4(t) = A\cos\left(2\pi f_c t + \frac{7\pi}{4}\right). \tag{4.5}$$

Applying the complex envelop representation, the mapping scheme can be alternatively specified as

$$00 \Longrightarrow s_1 = Ae^{j\frac{\pi}{4}};$$

$$01 \Longrightarrow s_2 = Ae^{j\frac{3\pi}{4}};$$

$$11 \Longrightarrow s_3 = Ae^{j\frac{5\pi}{4}};$$

$$10 \Longrightarrow s_4 = Ae^{j\frac{7\pi}{4}}. \tag{4.6}$$

The mapping scheme can also be specified in terms of constellation points in the signal space as shown in Figure 4.3. The coordinates of constellation points, also the output I/Q values, are $\pm A\cos(\pi/4) = \pm A/\sqrt{2}$. Note that the mapping scheme ensures that closest constellation points differ by only one bit, which minimizes the bit error probability for the same symbol error rate. Such mapping strategy is typically known as the *Gray coding* scheme.

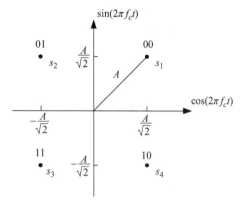

Figure 4.3   Constellation points of QPSK

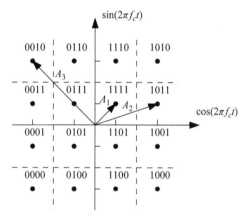

Figure 4.4   Sample constellation mapping for 16-QAM scheme

As another illustration, let us consider the more general QAM scheme, which is popular in high data rate wireless transmission systems. With M-ary QAM, total M sinusoidal with different amplitude and/or phases are used to carry bit sequences of length $\log_2 M$. The modulated symbols are of the general form, as given in (4.2)–(4.4). For example, the constellation structure of 16-ary square QAM and bit sequence to constellation point mapping are shown in Figure 4.4. Each modulated symbol carries four coded information bits with Gray coding. The modulated symbols have three possible magnitude A_1, A_2, and A_3. The coordinates of the constellation points are $\pm A_1/\sqrt{2}$ and $\pm A_2/\sqrt{2}$.

When the incoming coded data stream arrives at the modulator, the mapping scheme is sequentially applied. The transmitted signal of linear bandpass modulation schemes consist of a sequence of sinusoidal with different phase and/or amplitude. Mathematically, the modulated signal can be written as

$$s(t) = \sum_{i=-\infty}^{+\infty} s_i(t - iT_s) = \sum_{i=-\infty}^{+\infty} A_i g(t - iT_s) \cos(2\pi f_c(t - iT_s) + \theta_i), \qquad (4.7)$$

where $g(t)$ is the pulse shape, assumed to be unit rectangle pulse of duration T_s here for the sake of clarity. In practice, the raised cosine pulse and Gaussian pulse are widely used to achieve a suitable trade-off between bandwidth efficiency and performance. In the complex envelop format, $s(t)$ can be rewritten as

$$s(t) = \mathrm{Re} \left\{ \sum_{i=-\infty}^{+\infty} s_i \cdot g(t - iT_s) e^{j2\pi f_c(t - iT_s)} \right\}. \qquad (4.8)$$

Such digitally modulated signal will be transmitted from the transmit antenna after proper RF processing.

Example: Linear bandpass modulation

A bit sequence of 101001111000 is being transmitted using linear bandpass modulation schemes.

1. Determine the modulated symbols when the system uses the standard QPSK modulation scheme with bit pairs to symbol phase mapping scheme given in Figure 4.3.
2. What if the 16-ary QAM modulation scheme with mapping scheme given in Figure 4.4 is used instead.

Solutions:

1. Applying the mapping scheme for QPSK, we can determine the transmitted symbols in complex envelop form as

Bit pairs	10	10	01	11	10	00
Modulated symbols	$Ae^{j\frac{7\pi}{4}}$	$Ae^{j\frac{7\pi}{4}}$	$Ae^{j\frac{3\pi}{4}}$	$Ae^{j\frac{5\pi}{4}}$	$Ae^{j\frac{7\pi}{4}}$	$Ae^{j\frac{\pi}{4}}$

Figure 4.5(a) shows the resulting modulated signal assuming $f_c T_s = 2$.

2. With 16-QAM, the modulated symbols are

Bit sequences	1010	0111	1000
Modulated symbols	$A_3 e^{j\frac{\pi}{4}}$	$A_1 e^{j\frac{5\pi}{4}}$	$A_3 e^{j\frac{7\pi}{4}}$

Figure 4.5(b) shows the resulting modulated signal assuming $f_c T_s = 2$. Note that 16-QAM needs half symbol periods to transmit the same number of coded bits.

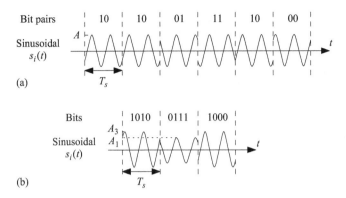

Figure 4.5   Modulated signal with (a) QPSK and (b) 16-QAM

4.3   Optimal detection in presence of AWGN

The objective of the demodulator is to arrive at the optimal detection decision about transmitted symbols based on the received signal. For linear modulation schemes, the modulated symbols reside in the two-dimensional space defined by $\cos(2\pi f_c t)$ and $\sin(2\pi f_c t)$. The optimal detection decision can be reached by applying the maximum likelihood principle within such space. Specifically, given the projection of the received signal in the signal space, the best estimate of the transmitted symbol would be the one most similar to the received signal. If the channel only introduces additive white Gaussian noise (AWGN), it can be verified that such approach will lead to the optimal detection performance for linear modulation schemes.

To implement such a detection principle, the demodulator first needs to determine the projection of the received signal in the space defined by $\cos(2\pi f_c t)$ and $\sin(2\pi f_c t)$. One common approach is to use *matched filter*. Figure 4.6 shows the structure of a matched filter-based demodulator for digital bandpass modulation schemes. In particular, the demodulator first multiplies locally generated carrier to the received signal and then applies matched filter. The matched filter matches to the shaping filter at the transmitter. If $g(t)$ is used as shaping pulse, then the matched filter response should be $g(T_s - t)$. As such, the output of the in-phase branch over a symbol period is given by

$$\hat{s}_I = \int_0^{T_s} r(t)g(T_s - t)\cos(2\pi f_c t)dt, \tag{4.9}$$

and that of the quadrature branch given by

$$\hat{s}_Q = \int_0^{T_s} r(t)g(T_s - t)\sin(2\pi f_c t)dt. \tag{4.10}$$

The projection of the received signal in the signal space, also referred to as the received symbol, is given by $r_i \triangleq \hat{s}_I + j\hat{s}_Q$. Applying the maximum likelihood principle, the demodulator will detect the constellation symbol closest to r_i as the transmitted symbol and output the corresponding bit sequence.

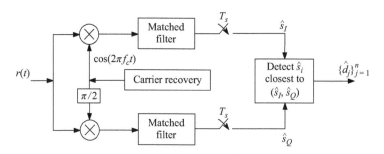

Figure 4.6   Structure of linear demodulator based on matched filter

When operating over AWGN channel, the received signal is given by

$$r(t) = s(t) + n(t), \tag{4.11}$$

where $n(t)$ is the additive white Gaussian noise with zero mean and power spectral density $N_0/2$. The in-phase component of the received symbol $\hat{s}_I$ is given by

$$\hat{s}_I = \int_0^{T_s} s(t)g(T_s - t) \cos(2\pi f_c t)dt + n_I, \tag{4.12}$$

where $n_I = \int_0^{T_s} n(t)g(T_s - t) \cos(2\pi f_c t)dt$ is the projection of noise signal to the in-phase direction. Under the assumption of perfect phase recovery and synchronization, we can show that

$$\int_0^{T_s} s(t)g^*(T_s - t) \cos(2\pi f_c t)dt = A_i \cos \theta_i. \tag{4.13}$$

As such, the in-phase component of the received symbol is the sum of the in-phase component of the transmitted symbol and the projection of $n(t)$ in the in-phase direction. Similarly, we can show that the quadrature component $\hat{s}_Q$ is the sum of the quadrature component of the transmitted symbol and the projection of $n(t)$ in the quadrature direction, i.e.,

$$\hat{s}_Q = A_i \sin \theta_i + n_Q, \tag{4.14}$$

where $n_Q = \int_0^{T_s} n(t)g(T_s - t) \sin(2\pi f_c t)dt$.

It follows that the received symbol in the signal space can be written as

$$r_i = s_i + n_i, \tag{4.15}$$

where $s_i = A_i e^{j\theta_i}$ is the transmitted symbol and $n_i = n_I + jn_Q$ is the projection of the noise signal $n(t)$ in the signal space. The other components of $n(t)$ are orthogonal to the signal space and, therefore, have no effect on detection decision. For the same reason, r_i is usually referred to as the *sufficient statistics* for the detection of linear modulation schemes. With the addition of n_i, r_i will be different from the transmitted symbol s_i. It can be shown that for AWGN channel, n_I and n_Q are independent Gaussian random variables with zero mean. As such, n_i is more likely to have small magnitude. If s_i is transmitted, r_i is more likely to be close to s_i in the signal space. Therefore, detecting the constellation symbol closest to r_i as the transmitted one is the optimal detection strategy for AWGN channel in the maximum likelihood sense.

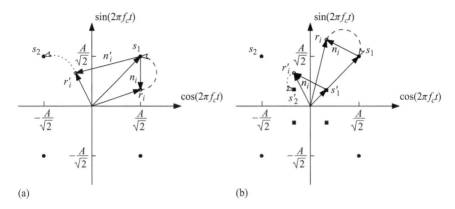

Figure 4.7   Detection error over AWGN channel: (a) effect of noise magnitude and (b) effect of symbol magnitude

With the application of the above detection principle, the signal space of linear modulation schemes can be divided into nonoverlapping *decision regions* for each constellation symbols. The boundaries of the decision regions for QPSK constellation shown in Figure 4.3 are the two axes, whereas that for 16-QAM constellation are marked as dashed lines Figure 4.4. Symbol detection error occurs when symbol s_i was transmitted but the received symbol r_i falls into the decision region of other constellation symbols. Such event may be the result of relatively large noise projection n_i and/or relatively small separation between possible transmitted symbols. n_i with large magnitude can cause that the received symbol r_i enters the decision region of other symbol than the transmitted symbols. Figure 4.7 illustrates the detection error for QPSK due to noise effect. We can see from Figure 4.7(a) that n_i with smaller magnitude n_i will not lead to detection error, whereas larger noise vector n_i' will result in symbol s_2 detected while s_1 was transmitted. On the other hand, the probability of symbol detection error will be reduced if the constellation symbols are further apart from each other in the signal space. We can see from Figure 4.7(b) that n_i will not cause detection error if the distance between symbols is relatively large. When the distances reduce, the same n_i will lead to detection error, i.e., s_2' is detected while s_1' is transmitted. Note that for linear bandpass modulation, the distance between modulated symbols in the signal space is proportional to the symbol magnitude. Smaller symbol magnitude implies smaller distance between modulated symbols.

4.4   Detection performance over AWGN channel

The distance between modulated symbols of general linear modulation schemes in the signal space is proportional to the average symbol energy, or equivalently, the transmitted signal power. The average magnitude of noise projection in the signal space is proportional to the noise power. It follows that the detection performance of digital linear modulation schemes over AWGN channels depends on the relative magnitude of transmitted signal power and noise power. The figure of merit that quantifies the quality of an AWGN channel is the received signal-to-noise ratio (SNR), defined as

$$\text{SNR} = \frac{P_s}{N}, \tag{4.16}$$

where P_s is the power of the transmitted signal $s(t)$, and N is the noise power. Assuming bandpass transmission with single side bandwidth B_s, the noise power can be determined as

$$N = N_0/2 \cdot 2B_s = N_0 B_s. \tag{4.17}$$

For digital linear modulation schemes under consideration, the power of the transmitted signal $s(t)$ depends on average energy per modulated symbol, E_s, and symbol duration, T_s. Specifically, we have

$$P_s = E_s/T_s. \tag{4.18}$$

It follows, the receiver SNR over AWGN channel is given by

$$\text{SNR} = \frac{E_s}{N_0 B_s T_s}. \tag{4.19}$$

The product of $B_s T_s$ typically equal to a constant, depending on the adopted pulse-shaping functions [1]. Therefore, the figure of merit that measures the quality of the AWGN channel is $\gamma_s = E_s/N_0$, which is typically referred to as SNR per symbol.

The modulated symbols of M-ary linear modulation schemes are in general given by

$$s_i(t) = A_i \cos(2\pi f_c t + \theta_i), \quad 0 \le t \le T_s, \ i = 1, 2, \ldots, M. \tag{4.20}$$

Note that we assume the unit rectangular pulse shape of duration T_s. The energy of the ith symbol can be calculated as

$$E_i = \int_0^{T_s} s_i^2(t) dt = \frac{A_i^2 T_s}{2}. \tag{4.21}$$

Therefore, the average symbol energy is calculated as

$$E_s = \frac{1}{M} \sum_i E_i. \tag{4.22}$$

For MPSK schemes, all M modulated symbols have the same energy $E_s = \frac{A^2 T_s}{2}$, where A is the common amplitude of the symbols. We can see that the transmission power is proportional to A^2, as expected.

Example: Average symbol energy of 16-QAM

Determine the average symbol energy of square 16-QAM scheme with constellation structure showing in Figure 4.4, in terms of the minimum distance between constellation points, denoted by $d_{\min}$.

Solutions: For 16-QAM, the coordinates of constellation points, also the output values of bits-to-I/Q mapper, are $\pm d_{min}/2$ and $\pm 3d_{min}/2$. As such, the three possible magnitude of the modulated symbols are related to d_{min} as

$$A_1 = \sqrt{(d_{min}/)2^2 + (d_{min}/2)^2} = d_{min}/\sqrt{2}, \tag{4.23}$$

$$A_2 = \sqrt{(d_{min}/)2^2 + (3d_{min}/2)^2} = \sqrt{5}d_{min}/\sqrt{2}, \tag{4.24}$$

and

$$A_3 = \sqrt{(3d_{min}/)2^2 + (3d_{min}/2)^2} = 3d_{min}/\sqrt{2}. \tag{4.25}$$

Out of 16 constellation symbols, 4 symbols have energy of $d_{min}^2 T_s/4$, 8 have energy of $5d_{min}^2 T_s/4$, and 4 have energy of $9d_{min}^2 T_s/4$. Finally, the average symbol energy of 16-QAM is calculated as

$$E_s = \frac{1}{16}\left(4 \cdot d_{min}^2 T_s/4 + 8 \cdot 5d_{min}^2 T_s/4 + 4 \cdot 9d_{min}^2 T_s/4\right) = \frac{5}{4}d_{min}^2 T_s. \tag{4.26}$$

The average symbol energy E_s, and also the transmitted signal power P_s, is proportional to d_{min}^2.

The detection performance of digital modulation schemes over AWGN channel depends on relative magnitude of E_s and N_0, or equivalently the SNR per symbol $\gamma_s = E_s/N_0$. In particular, the probability of bit error for binary PSK (BPSK) modulation with coherence detection over AWGN channel can be shown to be given by

$$P_b(E) = Q\left(\sqrt{2\gamma_s}\right), \tag{4.27}$$

where $Q(\cdot)$ is the Gaussian Q-function defined as

$$Q(x) = \frac{1}{\sqrt{2\pi}} \int_x^\infty e^{-y^2/2} dy. \tag{4.28}$$

Note that for binary modulation schemes, symbol energy E_s is equal to bit energy E_b and SNR per symbol γ_s is the same as SNR per bit, defined by $\gamma_b = E_b/N_0$. The probability of bit error for binary ASK modulation is given by

$$P_b(E) = Q\left(\sqrt{\gamma_b}\right). \tag{4.29}$$

We can see that binary ASK requires 3 dB higher SNR to achieve the same error performance as BPSK.

QPSK essentially consists of two BPSK modulation on in-phase and quadrature branches, respectively. The bit error probability of QPSK with Gray coding is the same as that of BPSK, i.e., $P_b(E) = Q\left(\sqrt{2\gamma_b}\right)$, where $\gamma_b = \gamma_s/2$ for QPSK.

The symbol error probability of QPSK can be derived, while noting that symbol error occurs when either bit is detected in error, as

$$P_s(E) = 1 - \left[1 - Q\left(\sqrt{2\gamma_b}\right)\right]^2 \approx 2Q\left(\sqrt{2\gamma_b}\right). \tag{4.30}$$

The closed-form expression for the symbol/bit error probability of general M-ary PSK (MPSK), $M > 4$, does not exist. Meanwhile, the symbol error probability of MPSK can be approximately calculated as

$$P_s(E) \approx 2Q\left(\sqrt{2\gamma_s}\sin(\pi/M)\right). \tag{4.31}$$

The exact symbol error probability of M-ary ASK (MASK) over AWGN channel can be shown to be given by

$$P_s(E) = \frac{2(M-1)}{M}Q\left(\sqrt{\frac{6\gamma_s}{M^2-1}}\right), \tag{4.32}$$

where γ_s is the average SNR per symbol.

Finally, we consider the square M-ary QAM (MQAM) scheme. Square MQAM essentially consists of two independent $\sqrt{M}$-ary ASK schemes on in-phase and quadrature branches, each with half of the average symbol energy. As such, the symbol error probability of square MQAM scheme can be determined as

$$P_s(E) = 1 - \left[1 - \frac{2\left(\sqrt{M}-1\right)}{\sqrt{M}}Q\left(\sqrt{\frac{3\gamma_s}{M-1}}\right)\right]^2. \tag{4.33}$$

The symbol error probability and bit error probability of square MQAM can also be approximately calculated with the following simplified expressions:

$$P_s(E) \approx 4Q\left(\sqrt{\frac{3\gamma_s}{M-1}}\right), \tag{4.34}$$

and

$$P_b(E) \approx \frac{1}{5}\exp\left(-\frac{3\gamma_s}{2(M-1)}\right), \tag{4.35}$$

respectively.

Example: Bit error probability of BPSK

Consider a digital transmission system using binary PSK (BPSK) modulation scheme. The transmission power is 30 mW and the bandwidth of the transmitted signal is 30 kHz. The system is operating over AWGN channel, where the additive white Gaussian noise at the receiver has a power spectrum density of 10^{-4} mW/Hz. Determine the bit error probability of the system assuming the system uses a pulse-shaping function that leads to $B_s T_s = 1.2$.

Solutions: The received signal-to-noise power ratio is calculated as

$$\text{SNR} = \frac{30 \times 10^{-3}}{10^{-7} \times 30 \times 10^{3}} = 10 = 10\,\text{dB}. \tag{4.36}$$

The bit error probability of BPSK is a function of SNR per symbol $\gamma_s = E_s/N_0$, which is related to the SNR as $\text{SNR} = \gamma_s/(B_s T_s)$. It follows that the SNR per symbol is equal to

$$\gamma_s = \frac{\text{SNR}}{B_s T_s} = 10/1.2 = 8.3. \tag{4.37}$$

As such, the bit error probability of BPSK over the AWGN channel is determined, with the help of Q-function table in the appendix, as

$$P_b(E) = Q\left(\sqrt{2\gamma_s}\right) = 2.25 \times 10^{-5}. \tag{4.38}$$

4.5   Effect of flat fading channel phase

When the digital transmission system is operating over wireless channels, the transmitted signal, as given in (4.7), will experience fading effect. Let us consider slow frequency flat fading channel. The received bandpass signal is given by

$$r(t) = \text{Re}\left\{ z\left(\sum_{i=-\infty}^{+\infty} s_i g(t - iT_s) \right) e^{j2\pi f_c t} \right\} + n(t), \tag{4.39}$$

where z represents the complex channel gain, $g(t)$ is the pulse-shaping function, and $n(t)$ denotes the additive white Gaussian noise. The channel gain z is given by

$$z = \sum_{n=1}^{N} \alpha_n e^{-j\phi_n}, \tag{4.40}$$

where N is the number of propagation paths, α_n and ϕ_n are the amplitude and phase of the nth path, respectively. Under slow fading assumption, z remains constant over multiple symbol periods. The received baseband symbol over the ith symbol period after matched filter detection can be determined as

$$r_i = z \cdot s_i + n_i = z \cdot A_i e^{j\theta_i} + n_i, \tag{4.41}$$

where $s_i = A_i e^{j\theta_i}$ is the transmitted symbol and n_i is the projection of the AWGN noise in the signal space. Since channel gain z is complex, both the amplitude and phase of the transmitted symbol s_i will be affected by z. In particular, in absence of the noise term n_i, the received symbol amplitude will become $|z|A_i$ and the received symbol phase $\angle z + \theta_i$. In this and next section, we will examine the effects of the flat fading channel gain z on digital transmission and related countermeasures to mitigate these effects.

Let first consider the phase angle of the complex channel gain z, typically referred to as channel phase. Most linear modulation schemes, including PSK and QAM schemes, use the phase of the modulated symbols to carry information. The receiver will examine the phase of received symbol to determine what bits were transmitted over each symbol period. On the basis of the above input/output relationship over a symbol period, the received symbol phase will be $\angle z + \theta_i$ because of the introduction of the complex channel gain z, if we ignore the noise component n_i. The channel phase $\angle z$ may take any value between 0 and 2π. As such, the symbol detection decision can be incorrect even if there is no noise in the system. If not properly mitigated, the channel phase will have detrimental effect on the detection performance of digital transmission schemes that use symbol phase to carry information.

Example: Detection error due to channel phase

Let us consider the transmission of a QPSK modulated signal over flat fading channel. The transmitted symbols are $Ae^{j\frac{7\pi}{4}}$, $Ae^{j\frac{7\pi}{4}}$, $Ae^{j\frac{3\pi}{4}}$, $Ae^{j\frac{5\pi}{4}}$, $Ae^{j\frac{7\pi}{4}}$, and $Ae^{j\frac{\pi}{4}}$. Let us assume that the complex channel gain during the transmission of first three symbols is equal to $0.01e^{j\pi/6}$ and then becomes $0.02e^{j\pi/3}$ when the last three symbols were transmitted. Neglecting noise and assuming the mapping scheme in Figure 4.3, determine the received symbol and the received bit sequence after maximum likelihood detection.

Solutions: Without the noise effect, the complex received symbols over flat fading channel are related to the transmitted symbol as

$$r_i = |z|Ae^{j\theta_i + j\angle z}. \tag{4.42}$$

On the basis of the given complex channel gains, the corresponding received symbols are $0.01Ae^{j\frac{7\pi}{4} + j\frac{\pi}{6}}$, $0.01Ae^{j\frac{7\pi}{4} + j\frac{\pi}{6}}$, $0.01Ae^{j\frac{3\pi}{4} + j\frac{\pi}{6}}$, $0.02Ae^{j\frac{5\pi}{4} + j\frac{\pi}{3}}$, $0.02Ae^{j\frac{7\pi}{4} + j\frac{\pi}{3}}$, and $0.02Ae^{j\frac{\pi}{4} + j\frac{\pi}{3}}$. The received symbol phases are determined as $\frac{7\pi}{4} + \frac{\pi}{6}$, $\frac{7\pi}{4} + \frac{\pi}{6}$, $\frac{3\pi}{4} + \frac{\pi}{6}$, $\frac{5\pi}{4} + \frac{\pi}{3}$, $\frac{7\pi}{4} + \frac{\pi}{3}$, and $\frac{\pi}{4} + \frac{\pi}{3}$.

The transmitted symbol can be determined according to the maximum likelihood principle. The detected symbols and the corresponding bit pairs are obtained as follows:

Detected symbols	$Ae^{j7\pi/4}$	$Ae^{j7\pi/4}$	$Ae^{j3\pi/4}$	$Ae^{j7\pi/4}$	$Ae^{j\pi/4}$	$Ae^{j3\pi/4}$
Bit pairs	10	10	01	10	00	01

We can see that 3 out of 12 transmitted bits are detected in error due to the channel phase, even in a noise-free scenario.

There are two major approaches to mitigate the effect of channel phase on digital wireless transmission, namely (i) channel estimation/phase correction and (ii) differential modulation, both of which are further explained in the following subsections.

4.5.1 Channel estimation

Channel estimation is the most intuitive approach to mitigate the effect of flat fading channel phase. In particular, if the receiver knows the complex channel gain z, or even only the channel phase $\angle z$, the receiver can remove the effect of channel phase on the received symbols by subtracting $\angle z$ from the received symbol phase. Referring to the previous example, if the receiver somehow acquires the knowledge of channel phase, i.e., $\pi/6$ for the first three symbols and $\pi/3$ for the last three symbols, the receiver may subtract $\pi/6$ from the phase value of first three received symbols and $\pi/3$ from the phase value of last three received symbols and then perform detection. Symbol detection will be always correct in noise-free environment. Such channel phase correction will still be effective even if a certain amount estimation error (less than $\pi/4$ for the QPSK example here) exist.

Many channel estimation/phase recovery solutions have been developed for digital transmission over AWGN channels and may be applicable to wireless channels. Depending on whether a pilot signal known to the receiver is transmitted or not, these solutions can be in general classified into two categories. When the transmitter sends pilot signal to facilitate channel estimation, the receiver may adopt a decision-directed channel estimation approach. One decision-directed technique uses data decision to remove the modulation effect in the received signal and feed the resulting signal to a phase locked loop (PLL) for phase estimation. When such pilot signal are not available at the receiver, the receiver will have to extract channel information from the modulated signal directly. Some estimation techniques typically assume certain distribution on transmitted symbols, with Gaussian assumption being the most common. Another approach is to apply nonlinear processing on the received signal to remove the modulation effects and apply again to a PLL. Different channel estimation techniques typically lead to different trade-offs of performance vs complexity.

Channel estimation for wireless fading channel is more challenging due to its time-varying nature. The channel estimates need to be periodically, if not continuously updated with period in the order of channel coherence time. For better detection performance, most advanced wireless transmission system adopts the pilot-symbol-assisted channel estimation approach. The basic idea is to transmit certain pilot symbols in a periodic fashion to the receiver. With the knowledge of the pilot symbols and their transmission instants, the receiver can effectively estimate the complex channel gain, including the channel phase. Let p denotes the complex pilot symbol that is transmitted. The received symbol during pilot transmission is given by

$$r_0 = z \cdot p + n_0, \tag{4.43}$$

where n_0 is the additive noise sample and z is the complex channel gain. An intuitive estimation of z would be

$$\hat{z} = \frac{r_0}{p}, \tag{4.44}$$

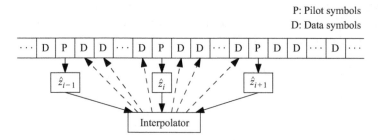

Figure 4.8   Channel estimation and interpolation using pilot symbols

which is actually the optimal estimate of z if the noise is Gaussian. The estimation accuracy can be improved if multiple pilot symbols, $p_1, p_2, \ldots, p_N$, are transmitted. The resulting channel estimate becomes

$$\hat{z} = \frac{1}{N} \sum_{i=1}^{N} \frac{r_i}{p_i}, \tag{4.45}$$

where r_i is the received symbol corresponding to pilot p_i. These estimates apply directly to those time instants when the pilot symbols were transmitted. When the channel introduces slow fading, the same estimate remains accurate for the duration of the order of a channel coherence time. Therefore, pilot symbols should be transmitted periodically with period in the order of a channel coherence time. To further improve the estimation accuracy during information data transmission, various interpolation algorithms can apply to consolidate the estimation results at pilot transmission instants. Of course, more frequent transmission of pilot symbols lead to better estimation accuracy, at the cost of lower information data transmission rate. Figure 4.8 illustrates the process of periodic pilot transmission as well as the interpolation process. From the complex channel gain estimate $\hat{z}$, we can easily obtain the channel phase estimate $\angle\hat{z}$ (as well as the channel amplitude estimate $|\hat{z}|$), which can be used to correct the channel phase shift affecting the transmitted signal.

4.5.2   Differential modulation

Another approach to mitigate the effect of channel phase is to apply differential modulation. Conventional modulation schemes map binary data information to the absolute phase value of modulated symbols, which will be affected by the flat fading channel gain during transmission. The effect of the channel is not reversible unless the receiver knows the channel gain. On the other hand, if the channel response remains the same over multiple symbol periods, as in slow fading environment, successive data symbols will be similarly affected by the channel. While the absolute phases of modulated symbols are changed by the channel, the relative phase shifts between successive received symbols remain the same. On the basis of these observations, differential modulation schemes were developed to overcome the channel phase effect. The basic idea of differential modulation is to map binary data information to the phase shifts between successive data symbols. The receiver can recover the transmitted data information by examining the phase shifts between successive received symbols.

We demonstrate the basic idea of differential modulation using the differential QPSK (DQPSK) modulation scheme, a variant of which is widely adopted in cellular wireless systems. Similarly to conventional QPSK modulation scheme, the transmitted symbols of DQPSK scheme over different symbol periods have the same amplitude and four different phase values, i.e., $\pi/4$, $3\pi/4$, $5\pi/4$, and $7\pi/4$. The possible phase changes between two consecutive symbols are 0, $\pi/2$, $-\pi/2$, and π. DQPSK modulation essentially maps information bit pairs to phase changes between successive symbols. The actual phases of modulated symbols will be determined using the phase of previous symbol as reference. As such, the transmission with differential modulation schemes always starts with a reference symbol transmission, whose phase is known to the receiver.

Example: Differential QPSK modulation

Let us consider a digital transmission system using differential QPSK modulation scheme. The bit pairs to phase change mapping scheme is specified as

$$00 \Longrightarrow 0,$$
$$01 \Longrightarrow \pi/2,$$
$$11 \Longrightarrow \pi,$$
$$10 \Longrightarrow 3\pi/2. \tag{4.46}$$

A bit sequence of 1001110001 is being transmitted over a fading channel. Determine the modulated symbols assuming that the phase of reference symbol is $\pi/4$.

Solutions: The bit pairs are first mapped to phase changes between symbols based on the above mapping scheme. Then, the phase of the transmitted symbols are sequentially determined using the phase of previous symbol as "current symbol phase = previous symbol phase + phase change." Finally, the transmitted symbols are determined.

Bit pairs	10	01	11	00	01
Phase changes	$3\pi/2$	$\pi/2$	π	0	$\pi/2$
Symbol phases (Reference phase $\pi/4$)	$7\pi/4$	$\pi/4$	$5\pi/4$	$5\pi/4$	$\pi/4$
Transmitted symbols (Reference symbol $Ae^{j\pi/4}$)	$Ae^{j7\pi/4}$	$Ae^{j\pi/4}$	$Ae^{j5\pi/4}$	$Ae^{j5\pi/4}$	$Ae^{j\pi/4}$

4.5.3 π/4 Differential QPSK

Note that the transmitted symbols with DQPSK scheme look exactly the same as those for conventional QPSK modulation scheme. Meanwhile, with conventional QPSK and DQPSK schemes, successive modulation symbols may have identical phase. In this scenario, the transmitted signal appears as an unmodelated carrier to the receiver. The receiver will have difficulty in maintaining synchronization with the transmitter. We also observe that there will be a phase change of π between successive symbols with QPSK and DQPSK. Such large phase change will lead to relatively large envelop variation in the transmitted signal. Specifically, the amplitude of the modulated signal will become essentially zero at certain time instant. Large envelop variation in modulated signal will lead to stringent requirement to the power amplifier and reduce the power efficiency of the transmitter.

To eliminate the phase change of 0 and π between successive modulated symbols, a variant of DQPSK, termed as $\pi/4$ differential QPSK ($\pi/4$-DQPSK), was developed and widely used in digital wireless transmission systems. The main difference between $\pi/4$-DQPSK and DQPSK is the mapping scheme from bit pairs to phase changes. A sample mapping scheme for $\pi/4$-DQPSK is given by

$$00 \Longrightarrow \pi/4,$$
$$01 \Longrightarrow 3\pi/4,$$
$$11 \Longrightarrow -3\pi/4,$$
$$10 \Longrightarrow -\pi/4. \tag{4.47}$$

Note that with $\pi/4$-DQPSK, the largest phase change between successive symbols are $\pm 3\pi/4$, which leads to smaller envelop variation compared to DQPSK. The smallest phase change is $\pm \pi/4$, good for time synchronization. Both phase changes of 0 and π of DQPSK are eliminated.

Example: $\pi/4$-DQPSK modulation

Determine the modulated symbols with $\pi/4$-DQPSK for the bit sequence 1001110001 assuming the above bit pair to phase changes mapping scheme and reference symbol phase of $\pi/4$.

Solutions: Following the same steps as previous example, the transmitted symbols with $\pi/4$-DQPSK for the bit sequence 1001110001 can be determined as

Bit pairs	10	01	11	00	01
Phase changes	$-\pi/4$	$3\pi/4$	$-3\pi/4$	$\pi/4$	$3\pi/4$
Symbol phases (Reference phase $\pi/4$)	0	$3\pi/4$	0	$\pi/4$	π
Transmitted symbols (Reference symbol $Ae^{j\pi/4}$)	A	$Ae^{j3\pi/4}$	A	$Ae^{j\pi/4}$	$Ae^{j\pi}$

We can see that the modulated symbols with $\pi/4$-DQPSK take eight different possible phase values. In this particular example, the odd symbols have phase values of 0, π, and $\pm\pi/2$, whereas the even symbols $\pi/4$, $3\pi/4$, $5\pi/4$, and $7\pi/4$. The transmission seems to be switching between two QPSK constellation structures, where one constellation is the obtained by shifting the other one by $\pi/4$, hence the name of $\pi/4$-DQPSK.

When the modulated signal is transmitted through fading channels, the phase of the transmitted symbols are modified by the channel. The received symbol phases become the transmitted symbol phases plus the channel phase. Meanwhile, since the information is carried by the phase changes between successive symbols with differential modulation, the correct phase difference can be determined at the receiver if the channel phase remains the same for successive symbols. For slow fading environment, the symbol period is much smaller than the channel coherence time. A large number of transmitted symbols will be affected by identical channel response. Even if the channel phase changes slightly between successive symbols, the correct detection of phase changes is also feasible. Without the noise effect, the phase changes between successive symbols can be detected in error only when the channel phase dramatically changes between successive symbol periods.

Example: $\pi/4$-DQPSK demodulation

We consider the transmission of the modulated symbols for bit sequence 1001110001 in previous example over a flat fading channel. The effect of additive noise is neglected (we will examine the noise effect in the next subsection). Determine the received bit sequence for the following cases:

1. The complex channel gain remains $z = ce^{\pi/2}$ for the transmission duration of all symbols, including the reference symbols.
2. The channel phase is equal to $\pi/2$ during the transmission of initial reference and first two symbols and changes to $2\pi/3$ afterwards.

Solutions:

1. The received symbols are determined, applying the flat fading channel with negligible noise assumption, as

Received symbols (Reference $cAe^{j5\pi/4}$)	$cAe^{\pi/2}$	$cAe^{j7\pi/4}$	$cAe^{\pi/2}$	$cAe^{j3\pi/4}$	$cAe^{j3\pi/2}$

   The receiver can determine the phase change between successive symbols as "current phase–previous phase," which leads to

Phase changes	$-\pi/4$	$3\pi/4$	$-3\pi/4$	$\pi/4$	$3\pi/4$

   The information bit sequence 1001110001 can be perfectly recovered after performing reverse mapping from phase changes to bit pairs.

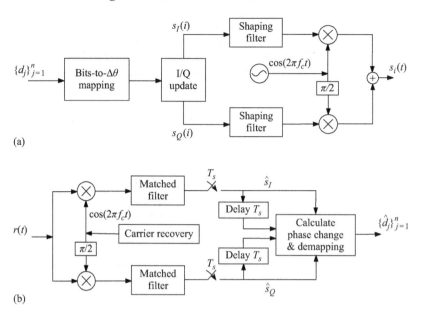

(a)

(b)

Figure 4.9   Structure of differential (a) modulator and (b) demodulator

2. Now that the channel phase changes during the transmission of the modulated symbols. The received symbols are determined as

---

Received symbols
(Reference $cAe^{j5\pi/4}$) $cAe^{\pi/2}$ $cAe^{j7\pi/4}$ $cAe^{2\pi/3}$ $cAe^{j11\pi/12}$ $cAe^{j5\pi/3}$

The phase changes between successive symbols can be calculated as

---

Phase changes $-\pi/4$ $3\pi/4$ $11\pi/12$ $\pi/4$ $3\pi/4$

After performing the reverse mapping from phase changes to bit pairs, the bit sequence is determined as 1001010001. The third bit pair is detected to 01 in error because $11\pi/12$ is closer to $3\pi/4$ than $-3\pi/4$. The detection of other bit pairs are not affected by the change in channel phase. Such detection error occurs only when the channel phase between successive symbols changes dramatically. In this particular example, the change needs to be greater than $\pi/4$, which typically will not happen frequently in slow fading environment.

The generic modulator and demodulator structure of differential PSK (DPSK) schemes are shown in Figure 4.9. At the modulator, the phase change will be used to update the in-phase and quadrature components for the current symbol period. The demodulator compares the phases of received symbols over successive symbol periods to determine the transmitted bit sequences. Such generic structure can be further specialized for specific DPSK schemes.

4.6 Effect of flat fading channel magnitude

In this section, we examine the effect of the amplitude of flat fading channel gain $|z|$ on digital wireless transmission. For the sake of clarity, we assume that either the channel phase has been accurately estimated and properly compensated or the system adopts certain form of differential modulation scheme to eliminate the effect of channel phase. With such assumption, the complex baseband received symbol over a symbol period becomes

$$r_i = |z|A_i e^{j\theta_i} + n_i. \tag{4.48}$$

The receiver needs to detect the transmitted symbol based on r_i. The optimal decision rule will detect the constellation symbol among all constellation symbols that is the closest to r_i as the transmitted symbol.

We can see that the channel amplitude affects the magnitude of received symbol. In general, $|z|$ is much smaller than 1 due to various propagation effects. Meanwhile, for the same transmitted symbol energy and noise power, the smaller the channel magnitude $|z|$, the larger the probability of detection error. Specifically, smaller $|z|$ will lead to smaller received symbol magnitude. A noise vector realization with small magnitude may cause the received symbol to be more similar to other constellation symbols than the transmitted one, resulting detection error. Figure 4.10 illustrates the effect of $|z|$ on the detection performance of QPSK modulation scheme. Note that here s_1 was transmitted. In the first case, the received symbol is $r_i = |z|s_1 + n_i$ with relatively large $|z|$. As the noise realization has small magnitude, the transmitted symbol s_1 can be correctly detected. In the second case, the received symbol is

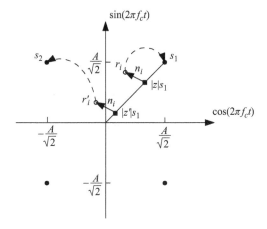

Figure 4.10   Effect of channel amplitude on detection performance

$r_i' = |z'|s_1 + n_i$ with relatively small $|z'|$. For the same noise realization, the detection is in error because of smaller received symbol magnitude.

Essentially, the channel amplitude $|z|$ affects the power of signal component in the received signal. The instantaneous power of signal component collected by the receiver is given by

$$P_r = |z|^2 P_s = |z|^2 \frac{E_s}{T_s}. \tag{4.49}$$

The square of the channel amplitude $|z|^2$ is usually referred to as *channel power gain*. It follows that the instantaneous received SNR per symbol becomes

$$\gamma_s = |z|^2 \frac{E_s}{N_0}, \tag{4.50}$$

which varies dramatically with instantaneous channel amplitude $|z|$. At a particular time instant, the wireless channel can be viewed as an AWGN channel with received SNR $|z|^2 \frac{E_s}{N_0}$. As such, the instantaneous bit error probability will depend on the realization of the complex channel gain $|z|$ as well as the adopted digital modulation scheme.

Example: Bit error probability over fading channels

Consider a digital transmission system operating over a wireless channel using BPSK modulation scheme. The transmission power is 30 mW, the bandwidth of the transmitted signal is 30 kHz, and the adopted pulse-shaping function leads to $B_s T_s = 1$. The additive white Gaussian noise at the receiver has a power spectrum density of 10^{-4} mW/Hz. Assume the channel amplitude of the wireless channel $|z|$ takes only two possible values. Specifically, $|z|$ is equal to 0.7 for 99% of the time and equal to 0.1 for 1% of the time. Determine the instantaneous error rate of the transmission.

Solutions: Noting that the transmit power $P_s = E_s/T_s = 30$ mW, it follows that the SNR per symbol at the transmitter is calculated as

$$\frac{E_s}{N_0} = \frac{P_s}{B_s N_0} = \frac{30 \times 10^{-3}}{30 \times 10^3 \times 10^{-7}} = 10. \tag{4.51}$$

When the channel amplitude $|z|$ is equal to 0.7, then the instantaneous received SNR per symbol γ_s is equal to $\gamma_s = 0.7^2 \cdot 10 = 4.9$. The corresponding instantaneous bit error probability for BPSK modulation is $Q(\sqrt{9.8}) = 8.2 \times 10^{-4}$. Meanwhile, when $|z| = 0.1$, we have $\gamma_s = 0.1$. The corresponding instantaneous bit error probability becomes 0.38.

4.6.1   Performance measures

We can see from the above example that the instantaneous bit error probability of digital transmission over wireless channel varies dramatically with the channel amplitude

$|z|$, or equivalently, the channel power gain $|z|^2$. As such, the instantaneous bit error rate cannot effectively characterize the transmission performance over fading channels. The performance measures that are commonly used for fading wireless channels include *outage probability* and *average error rate*.

Outage occurs when the instantaneous received signal power is too low for reliable information transmission. Over fading wireless channels, the system may experience outage even when the average received SNR is sufficiently high after considering path loss and shadowing effects. Note that the flat fading channel gain varies dramatically around its average due to the noncoherent addition of signal copies from different paths. The performance metric outage probability characterizes how often the instantaneous link quality becomes unacceptable. Outage probability, denoted by P_{out}, is defined as the probability that the instantaneous received SNR falls below a certain SNR threshold. Mathematically, P_{out} is given by

$$P_{out} = \Pr\left[\gamma_s < \gamma_{th}\right], \tag{4.52}$$

where γ_{th} is the SNR threshold. The outage threshold is usually chosen according to a particular reliability requirement. If the instantaneous bit error probability is expected to be less than BER_0, then γ_{th} should be chosen to be the smallest SNR value such that the bit error probability of the adopted modulation scheme is less than BER_0. Referring to the above BPSK example, if the maximum instantaneous tolerable error rate, BER_0, is 10^{-3}, then γ_{th} should be chosen as $(Q^{-1}(10^{-3}))^2/2$, where $Q^{-1}(\cdot)$ denotes the inverse of Gaussian Q-function. The resulting γ_{th} value is 4.77 or equivalently 6.79 dB. The outage probability for the particular wireless channel is 0.01 as γ_s is less than 4.77 for 1% of the time.

Average error rate, also referred to as the average probability of error, characterizes the average link quality of wireless fading channels. The average error rate is defined as the time average of the instantaneous error probability. Note that at any time instant, the fading channel can be viewed as an AWGN channel with received SNR equal to $\gamma_s = |z|^2 \frac{E_s}{N_0}$. Therefore, the average error rate of a modulation scheme over flat fading channel can be calculated by averaging the instantaneous error rate, which is the error probability of this modulation scheme over AWGN channel with received SNR value γ_s over the statistical distribution of γ_s.

Example: Average bit error rate of BPSK

Consider the digital transmission with BPSK over the special wireless channel, where the received SNR γ_s is equal to 4.9 for 99% of the time and equal to 0.1 for 1% of the time. Determine the average bit error rate of the transmission system.

Solutions: Note that the instantaneous received SNR of 4.9 leads to the error probability of 8.2×10^{-4} for BPSK and that of 0.1 leads to the error probability of 0.38. The system enjoys the low error probability 8.2×10^{-4} for

99% of the time. Meanwhile, the average bit error rate of the transmission system can be evaluated as

$$\overline{P}_E = 99\% \cdot 8.2 \times 10^{-4} + 1\% \cdot 0.38 = 3.9 \times 10^{-3}. \tag{4.53}$$

We can observe that the bad channel condition, although happening with very small probability, dominate the average error rate performance of the system. This observation holds in general for digital transmission over fading channels, as the contribution of good channel condition to the average error rate is relatively small. Therefore, the effective ways to improve the performance include (i) reducing the occurrence of bad channel conditions and (ii) improving the quality of bad channels. For example, if we can reduce the time percentage of bad channel to 0.1% and increase that of good channel to 99.9%, then the average BER will be 1.11 × 10^{-3}. It is exactly from these perspective that diversity combining techniques, one of the most effective fading mitigation solution, try to improve the performance of digital transmission over fading channels. We will elaborate on diversity technique in later chapter.

The analysis of both outage probability and average error rate rely on the availability of the statistical characterization of the received SNR γ_s. In particular, should the cumulative distribution function (CDF) of the received SNR γ_s be available, we can calculate the outage probability by evaluating the CDF of γ_s at γ_{th}, i.e.,

$$P_{\text{out}} = F_{\gamma_s}(\gamma_{\text{th}}), \tag{4.54}$$

where $F_{\gamma_s}(\cdot)$ denotes the CDF of γ_s. In the following subsection, we develop the statistical models of γ_s for various flat fading environment to facilitate the performance analysis of digital transmission over fading channels.

4.6.2   Statistical models for flat fading channel gain

Statistical characterization of the received SNR over realistic propagation environment, in terms of the PDF and CDF of γ_s, is essential to the performance analysis of digital transmission over fading channels. For flat fading environment, the instantaneous received SNR γ_s is given by $|z|^2 E_s/N_0$, where $|z|$ is the amplitude of the complex channel gain. As such, we will first derive the statistical distributions of $|z|$ and then apply the transformation of random variables to obtain the distribution of the received SNR γ_s.

Recall that for flat fading environment, the instantaneous complex channel gain z is given by

$$z = \sum_{n=1}^{N} \alpha_n e^{-j\phi_n}, \tag{4.55}$$

where N is the number of paths, α_n and ϕ_n are the amplitude and phase of the nth path, respectively. We can rewrite z in real/imaginary part format as

$$z = z_I + jz_Q, \tag{4.56}$$

where the real part z_I and the imaginary part z_Q are given by

$$z_I = \sum_{n=0}^{N} \alpha_n \cos \phi_n, \qquad z_Q = \sum_{n=0}^{N} \alpha_n \sin \phi_n, \tag{4.57}$$

respectively. It follows that the channel amplitude $|z| = \sqrt{z_I^2 + z_Q^2}$. When the number of multipath components is large, both z_I and z_Q are the sum of a large number of real numbers. We can apply the central limit theorem (CLT) and model both z_I and z_Q as Gaussian random variables. Since ϕ_ns vary quickly over time and are independent of α_ns, z_I and z_Q can be shown to be independent random variables. As such, z can be modeled as complex Gaussian random variable with independent real and imaginary parts.

Starting from this basic result, we can arrive at the statistics of $|z|$ depending on whether a line-of-sight (LOS) component exists or not. When there is no LOS component, the complex Gaussian random variable z can be assumed to have zero mean. It follows that the channel amplitude $|z| = \sqrt{z_I^2 + z_Q^2}$ is Rayleigh distributed with distribution function

$$p_{|z|}(x) = \frac{x}{\sigma^2} \exp\left[-\frac{x^2}{2\sigma^2}\right], \quad x \geq 0, \tag{4.58}$$

whereas the channel phase $\theta = \arctan(z_Q/z_I)$ is uniformly distributed over $[0, 2\pi]$. As such, the flat fading wireless channel without LOS is typically referred to as the *Rayleigh fading* channel.

Applying the transformation of random variables, we can show that the channel power gain $|z|^2$ follows an exponential distribution with PDF given by

$$p_{|z|^2}(x) = \frac{1}{2\sigma^2} \exp\left[-\frac{x}{2\sigma^2}\right], \quad x \geq 0, \tag{4.59}$$

where $2\sigma^2 = \overline{P}_r$ is the average received signal power, depending upon the path loss and shadowing effects.

Example: Outage probability of Rayleigh fading channel

Consider the wireless transmission over a Rayleigh fading channel. The transmit power is 10 dBm. The average received signal power at the receiver after considering path loss and shadowing effect is -10 dBm. What is the probability that the instantaneous received signal power is less than -20 dBm due to fading?

Solutions: Given the relationship that $P_r = P_t \cdot |z|^2$, the probability that the received signal power is less than -20 dBm is equal to the probability that the channel power gain $|z|^2$ is less than -30 dB, which can be calculated as

$$\Pr\left[|z|^2 < -30 \text{ dB}\right] = \int_0^{0.001} p_{|z|^2}(x)dx = 1 - \exp\left(-\frac{0.001}{2\sigma^2}\right). \tag{4.60}$$

Noting that the average channel power gain $2\sigma^2 = \overline{P_r} - P_t = -20$ dB $= 0.01$, the probability that the instantaneous received signal power is less than -20 dBm can be calculated to be equal to 9.5%.

When the LOS component exists, the real and imaginary parts of complex channel gain, z_I and z_Q, are modeled as Gaussian random variables with nonzero mean. The value of the mean is related the gain of the LOS path. Starting from the joint PDF of z_I and z_Q, we can derive PDF of the instantaneous channel amplitude $|z| = \sqrt{z_I^2 + z_Q^2}$ as

$$p_{|z|}(x) = \frac{x}{\sigma^2} \exp\left[-\frac{x^2 + s^2}{2\sigma^2}\right] I_0\left(\frac{xs}{\sigma^2}\right). \tag{4.61}$$

where s^2 is the amplitude square of the LOS path gain, i.e., $s^2 = \alpha_0^2$, $2\sigma^2$ is the sum of average channel power gain of all non-LOS components, and $I_0(\cdot)$ is the first-kind modified Bessel function of zeroth order. Therefore, $|z|$ follows the Rician distribution. Alternatively, the Rician distribution function is given in terms of the so-called Rician fading parameter $K = s^2/2\sigma^2$ and the total average received power $\overline{P_r} = s^2 + 2\sigma^2$ as

$$p_{|z|}(x) = \frac{2(K+1)x}{\overline{P_r}} \exp\left[-K - \frac{(K+1)x^2}{\overline{P_r}}\right] I_0\left(2\sqrt{\frac{K(K+1)}{\overline{P_r}}}x\right). \tag{4.62}$$

The channel power gain $|z|^2$ for LOS scenario follows a noncentral χ^2 distribution with distribution function

$$p_{|z|^2}(x) = \frac{K+1}{\overline{P_r}} \exp\left[-K - \frac{(K+1)x}{\overline{P_r}}\right] I_0\left(2\sqrt{\frac{K(K+1)x}{\overline{P_r}}}\right). \tag{4.63}$$

Nakagami fading model is another statistical model for LOS fading scenario. This model was developed based on experimental measurements. With Nakagami model, the channel amplitude $|z|$ is modeled as a random variable with distribution function

$$p_{|z|}(x) = \frac{2m^m x^{2m-1}}{\Gamma(m)\overline{P_r}} \exp\left[-\frac{mx^2}{\overline{P_r}}\right], \tag{4.64}$$

where $\Gamma(\cdot)$ is the Gamma function and $m \geq 1/2$ is the Nakagami fading parameter. It follows that the distribution function of the channel power gain $|z|^2$ under Nakagami model is given by

$$p_{|z|^2}(x) = \left(\frac{m}{\overline{P}_r}\right)^m \frac{x^{m-1}}{\Gamma(m)} \exp\left[-\frac{mx}{\overline{P}_r}\right]. \tag{4.65}$$

With properly selected values for Nakagami parameter m, Nakagami model can apply to many fading scenarios. Specifically, when $m = 1$ (or $K = 0$ for Rician fading model), we have Rayleigh fading model. If m approaches ∞ (or K approaches ∞ for Rician model), then we arrive at no fading case. Nakagami model can approximate Rician fading well with $m = \frac{(K+1)^2}{2K+1}$. Finally, when $m < 1$, Nakagami model applies to more severe fading scenario than Rayleigh fading.

4.7   Performance analysis over fading channels

The received SNR γ_s is related to the channel amplitude as $\gamma_s = |z|^2 \frac{E_s}{N_0}$. Applying the function of random variable result, we can derive the distribution function of γ_s from those of $|z|$ and $|z|^2$. For example, the PDF of the received SNR over Rayleigh fading channel is given by

$$p_{\gamma_s}(\gamma) = \frac{1}{\overline{\gamma}_s} \exp\left(-\frac{\gamma}{\overline{\gamma}_s}\right), \quad \gamma \geq 0, \tag{4.66}$$

where $\overline{\gamma}_s = 2\sigma^2 \frac{E_s}{N_0}$ is the average received SNR. The PDF of received SNR over other fading channel models can be similarly obtained and summarized in Table 4.1. In Table 4.1, $\overline{\gamma}$ is the average received SNR, $\Gamma(\cdot,\cdot)$ is the incomplete Gamma function, and $Q_l(\cdot,\cdot)$ is the lth-order Marcum Q-function. We can readily analyze the performance of digital wireless communication systems over flat fading channels.

Table 4.1   Statistics of received SNR γ for three fading models under consideration

Model	Rayleigh	Rice	Nakagami-m
Parameter	$\cdot$	$K \geq 0$	$m \geq \frac{1}{2}$
PDF, $p_\gamma(x)$	$\frac{1}{\overline{\gamma}}e^{-\frac{x}{\overline{\gamma}}}$	$\frac{(1+K)}{\overline{\gamma}}e^{-K-\frac{1+K}{\overline{\gamma}}x}I_0\left(2\sqrt{\frac{1+K}{\overline{\gamma}}Kx}\right)$	$(\frac{m}{\overline{\gamma}})^m \frac{x^{m-1}}{\Gamma(m)}e^{-\frac{mx}{\overline{\gamma}}}$
CDF, $F_\gamma(x)$	$1-e^{-\frac{x}{\overline{\gamma}}}$	$1-Q_1\left(\sqrt{2K},\sqrt{\frac{2(1+K)}{\overline{\gamma}}x}\right)$	$1-\frac{\Gamma\left(m,\frac{m}{\overline{\gamma}}x\right)}{\Gamma(m)}$
MGF, $\mathcal{M}_\gamma(s)$	$(1-s\overline{\gamma})^{-1}$	$\frac{1+K}{1+K-s\overline{\gamma}}e^{\frac{s\overline{\gamma}K}{1+K-s\overline{\gamma}}}$	$\left(1-\frac{s\overline{\gamma}}{m}\right)^{-m}$

4.7.1 Outage analysis

Due to the fading effect, the instantaneous received SNR at the receiver may be too low to support a particular target BER. When the instantaneous received SNR γ_s is below the SNR threshold γ_{th}, the transmission link enters outage. The probability of outage, i.e., $P_{out} = \Pr[\gamma_s < \gamma_{th}]$, can be analytically calculated by evaluating the CDF of the received SNR, $F_{\gamma_s}(\cdot)$, γ_{th}. The CDF of γ_s for common fading channel models are also summarized in Table 4.1. As such, the outage probability can be readily evaluated. Specifically, for the Rayleigh fading case, the outage probability is given by

$$P_{out} = 1 - \exp\left(-\frac{\gamma_{th}}{\overline{\gamma}_s}\right). \tag{4.67}$$

Similarly, the outage probability of a point-to-point link under Rician fading model is given by

$$P_{out} = 1 - Q_1\left(\sqrt{2K}, \sqrt{\frac{2(1+K)\gamma_{th}}{\overline{\gamma}_s}}\right), \tag{4.68}$$

where $Q_1(\cdot, \cdot)$ is the Marcum Q-function. Finally, the outage probability under Nakagami fading is

$$P_{out} = 1 - \frac{\Gamma\left(m, \frac{m\gamma_{th}}{\overline{\gamma}_s}\right)}{\Gamma(m)}, \tag{4.69}$$

where $\Gamma(\cdot, \cdot)$ is the incomplete Gamma function.

Example: Required SNR for target outage requirement

Let us consider a digital transmission system using BPSK modulation scheme operating over Rayleigh fading channel. The transmission link quality is considered acceptable if the instantaneous BER is less than 10^{-4} for at least 99.5% of the time. Determine the minimum required average received SNR that the system needs to maintain.

Solutions: Noting that the instantaneous BER of BPSK modulation is related to the instantaneous SNR γ_s as $P_b(\gamma_s) = Q(\sqrt{2\gamma_s})$, where $Q(\cdot)$ is the Gaussian Q-function. Therefore, to satisfy the BER of 10^{-4}, the instantaneous SNR should be greater than $(Q^{-1}(10^{-4}))^2/2 = 7.08 = 8.5$ dB. The link reliability requirement can be mathematically expressed in terms of outage probability as

$$\Pr[\gamma_s < 8.5 \text{ dB}] < 0.5\%. \tag{4.70}$$

After applying the outage probability expression for Rayleigh fading channel while noting that $\gamma_{th} = 8.5\,\text{dB}$, we have

$$1 - \exp\left(-\frac{10^{0.85}}{\overline{\gamma}_s}\right) < 0.5\%. \tag{4.71}$$

Finally, solving for $\overline{\gamma}_s$, we can determine the minimum required value of $\overline{\gamma}_s$ as $31.4\,\text{dB}$.

4.7.2 PDF-based average error rate analysis

Average error rate can be calculated as the statistical average with respect to γ_s of the instantaneous error probability of digital modulation schemes. Given the fading channel model, we can evaluate the average error rate using the PDF of instantaneous SNR γ_s. In particular, the average error rate, denoted by $\overline{P}_E$, is calculated as

$$\overline{P}_E = \int_0^\infty P_E(\gamma) p_{\gamma_s}(\gamma) d\gamma, \tag{4.72}$$

where $P_E(\gamma)$ is the instantaneous error rate of the modulation scheme under consideration with instantaneous SNR γ and $p_{\gamma_s}(\gamma)$ is the PDF of γ_s. The instantaneous error rate $P_E(\gamma)$ is essentially the error probability of the chosen modulation scheme over AWGN channel with received SNR γ.

As an illustrative example, let us consider the average error rate performance of BPSK modulation scheme over Rayleigh fading channels. The instantaneous error rate of BPSK with instantaneous SNR γ is equal to $Q\left(\sqrt{2\gamma}\right)$. Under Rayleigh fading model, the PDF of the received SNR γ_s was given in (4.66). Therefore, the average error rate of BPSK over Rayleigh fading can be calculated as

$$\overline{P}_E = \int_0^\infty Q\left(\sqrt{2\gamma}\right) \frac{1}{\overline{\gamma}_s} \exp\left(-\frac{\gamma}{\overline{\gamma}_s}\right) d\gamma, \tag{4.73}$$

which can be simplified, after carrying out integration with some calculus manipulation, to the following closed-form result

$$\overline{P}_E = \frac{1}{2}\left(1 - \sqrt{\frac{\overline{\gamma}_s}{1 + \overline{\gamma}_s}}\right). \tag{4.74}$$

This general PDF-based approach applies to the average error rate analysis of other modulation schemes/fading channel models. We just need to plug in different distribution functions and/or instantaneous error rate expression and then perform the integration. It worths noting that in most cases, we cannot obtain a closed-form expression for the average error rate. For example, the average error rate of square M-QAM over Rician fading channel should be evaluated as

$$\overline{P}_E = \int_0^\infty \left\{ 1 - \left[1 - \frac{2(M-1)}{M} Q\left(\sqrt{\frac{3\gamma}{M^2 - 1}} \right) \right]^2 \right\}$$

$$\times \frac{K+1}{\overline{\gamma}_s} \exp\left[-K - \frac{(K+1)\gamma}{\overline{\gamma}_s} \right] I_0 \left(2\sqrt{\frac{K(K+1)\gamma}{\overline{\gamma}_s}} \right) d\gamma. \quad (4.75)$$

Because of the complex instantaneous error rate and PDF expressions involved, the average error rate for such case has to be evaluated through numerical method. Since the integration is carried over the whole real axis, it is challenging to obtain very accurate result with truncation followed by numerical integration. Alternatively, we can analyze the average error rate following a moment generating function (MGF)-based approach.

4.7.3   *MGF-based average error rate analysis*

An alternative analytical framework based on the MGF of random variables was developed to facilitate the accurate evaluation of the average error rate of digital transmission schemes over general fading channels. The MGF of a nonnegative random variable γ is defined as

$$\mathcal{M}_\gamma(s) = \int_0^\infty p_\gamma(\gamma) e^{s\gamma} d\gamma. \quad (4.76)$$

where s is a complex dummy variable and $p_\gamma(\cdot)$ is the PDF of γ. MGF is so named because it allows for easy calculation of the moments. Specifically, the nth moment of a random variable can be calculated as

$$\mathbf{E}[\gamma^n] = \frac{d^n}{ds^n} \mathcal{M}_\gamma(s) \bigg|_{s=0}. \quad (4.77)$$

The MGF of the received SNR for most popular fading channel models are readily available in a compact closed form. For example, the MGF of received SNR for Rayleigh fading case can be shown to be given by

$$\mathcal{M}_{\gamma_s}(s) = \left(1 - s\overline{\gamma}_s \right)^{-1}. \quad (4.78)$$

Those for Rician and Nakagami fading models are also available and presented in Table 4.1 for easy reference. These MGF results can help evaluate the average error rate performance of digital transmission scheme over fading channels. In particular, if the instantaneous error rate of a modulation scheme is of the form $P_E(\gamma) = a \exp(-b\gamma)$, where γ is the instantaneous received SNR, then the average error rate can be evaluated as

$$\overline{P}_E = \int_0^\infty a\exp(-b\gamma) p_{\gamma_s}(\gamma) d\gamma, \quad (4.79)$$

where $p_{\gamma_s}(\cdot)$ denotes the PDF of the received SNR γ_s. Applying the definition of the MGF, we can calculate the average error rate in terms of the MGF $\mathcal{M}_{\gamma_s}(s)$ as

$$\overline{P}_E = a\mathcal{M}_{\gamma_s}(-b). \tag{4.80}$$

For example, the instantaneous error rate of binary differential PSK (BDPSK) modulation scheme is given by $P_E(\gamma) = \exp(-\gamma)/2$. Applying the MGF approach, the average error rate of BDPSK over Rayleigh fading channels can be shown to be given by

$$\overline{P}_E = \frac{1}{2}\mathcal{M}_{\gamma_s}(-1) = \frac{1}{2(1+\overline{\gamma}_s)}, \tag{4.81}$$

where $\overline{\gamma}_s$ is the average received SNR.

The instantaneous error rate expression of many linear digital modulation schemes, including BPSK and square M-QAM, involves the Gaussian Q-function and its square. To apply the MGF to evaluate the average error rate, we need to evoke the alternative expression of the Gaussian Q-function and its square obtained by Craig in the early 1990s. Let us first consider digital modulation schemes whose instantaneous error rate expression is of the form $aQ\left(\sqrt{b\gamma}\right)$, where a and b are constants. The average error rate of such modulation scheme should be conventionally evaluated as

$$\overline{P}_E = \int_0^\infty aQ\left(\sqrt{b\gamma}\right)p_{\gamma_s}(\gamma)d\gamma. \tag{4.82}$$

The direct evaluation of the above expression after substituting in (4.28) will be very challenging, except for few special cases. Fortunately, the Gaussian Q-function can be alternatively given by

$$Q(x) = \frac{1}{\pi}\int_0^{\pi/2}\exp\left(\frac{-x^2}{2\sin^2\phi}\right)d\phi, \quad x > 0. \tag{4.83}$$

Substituting in the alternative expression of the Gaussian Q-function, we arrive at

$$\overline{P}_E = \int_0^\infty \frac{a}{\pi}\left(\int_0^{\pi/2}\exp\left(\frac{-b\gamma}{2\sin^2\phi}\right)d\phi\right)p_{\gamma_s}(\gamma)d\gamma. \tag{4.84}$$

After changing the order of integration and applying the definition of MGF, we can rewrite the average error rate as

$$\overline{P}_E = \frac{a}{\pi}\int_0^{\pi/2}\mathcal{M}_{\gamma_s}\left(\frac{-b}{2\sin^2\phi}\right)d\phi, \tag{4.85}$$

where $\mathcal{M}_{\gamma_s}(\cdot)$ is the MGF of the received SNR for the fading model under consideration. Note that the resulting expression only involves a single integration with respect to ϕ over the integral of $[0, \pi/2]$, which is much easier to evaluate numerically. For modulation schemes whose instantaneous error rate involve the square of Gaussian Q-function $Q^2(\cdot)$, we can use the following alternative equivalent expression in average error rate analysis

$$Q^2(x) = \frac{1}{\pi}\int_0^{\pi/4}\exp\left(\frac{-x^2}{2\sin^2\phi}\right)d\phi, \quad x > 0. \tag{4.86}$$

Example: Average error rate analysis

Apply the MGF approach to derive the average error rate expression for square M-QAM modulation over Nakagami fading channels.

Solutions: The instantaneous BER of square M-QAM modulation is related to the instantaneous SNR γ_s as

$$P_E(\gamma_s) = \frac{4(\sqrt{M}-1)}{\sqrt{M}} Q\left(\sqrt{\frac{3\gamma}{M-1}}\right) - \left(\frac{2(\sqrt{M}-1)}{\sqrt{M}}\right)^2 Q\left(\sqrt{\frac{3\gamma}{M-1}}\right)^2, \quad (4.87)$$

where $Q(\,\cdot\,)$ is the Gaussian Q-function. Substituting into (4.72) with the alternative expression of $Q(\,\cdot\,)$ and $Q^2(\,\cdot\,)$, we arrive at

$$\overline{P}_E = \int_0^\infty \frac{4\left(\sqrt{M}-1\right)}{\sqrt{M}\pi} \left(\int_0^{\pi/2} \exp\left(\frac{-3\gamma}{(M-1)2\sin^2\phi}\right) d\phi\right) p_{\gamma_s}(\gamma) d\gamma$$

$$- \int_0^\infty \frac{4\left(\sqrt{M}-1\right)^2}{M\pi} \left(\int_0^{\pi/4} \exp\left(\frac{-3\gamma}{(M-1)2\sin^2\phi}\right) d\phi\right) p_{\gamma_s}(\gamma) d\gamma. \quad (4.88)$$

Applying the definition of MGF after changing the order of integration, we have

$$\overline{P}_E = \frac{4\left(\sqrt{M}-1\right)}{\sqrt{M}\pi} \int_0^{\pi/2} \mathcal{M}_{\gamma_s}\left(\frac{-3}{(M-1)2\sin^2\phi}\right) d\phi$$

$$- \frac{4\left(\sqrt{M}-1\right)^2}{M\pi} \int_0^{\pi/4} \mathcal{M}_{\gamma_s}\left(\frac{-3}{(M-1)2\sin^2\phi}\right) d\phi. \quad (4.89)$$

Finally, we arrived at the final expression of the average error rate after plugging-in the Nakagami MGF in Table 4.1 as

$$\overline{P}_E = \frac{4(\sqrt{M}-1)}{\sqrt{M}\pi} \int_0^{\pi/2} \left(1 - \frac{-3\overline{\gamma}_s}{2m(M-1)\sin^2\phi}\right)^m d\phi$$

$$- \frac{4(\sqrt{M}-1)^2}{M\pi} \int_0^{\pi/4} \left(1 - \frac{-3\overline{\gamma}_s}{2m(M-1)\sin^2\phi}\right)^m d\phi, \quad (4.90)$$

where m is the Nakagami fading parameter and $\overline{\gamma}_s$ is the average received SNR.

Figure 4.11 plots the average error rate of BPSK and binary DPSK over fading channels as the function of the average received SNR. For reference, the error rate of these modulation schemes over AWGN channel (with received SNR equal to the average SNR) is also plotted on the same figure. As we can see, the error performance

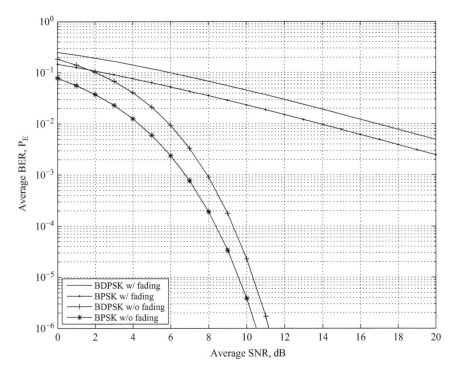

Figure 4.11   Error rate comparison between fading and nonfading environments

degrades dramatically over fading channels. In particular, the error rate decreases in an exponential rate over AWGN channel as the SNR increases but in a linear rate for the fading channel case. As such, to achieve the same average error rate over fading channels, we need to maintain much higher average SNR. An intuitive explanation of this phenomenon is that fading causes the occurrence of very low received signal power due to destructive addition of different multipath signals, which is usually referred to as *deep fade*. These deep fade will lead to large instantaneous error rate and in turn deteriorate the average error rate performance. Various wireless transmission technologies have been proposed to improve the performance of digital transmission over fading channels and will be presented in the following chapters.

4.8   Further readings

Our discussion here is limited to linear bandpass modulation schemes and phase differential modulation schemes. Further discussion about other digital bandpass modulation schemes and more detailed explanation on their performance over AWGN channel can be found in [1, Chapters 4 and 5]. The spectrum analysis of various modulation schemes are available in [2,3]. Various synchronization and carrier phase recovery techniques are presented in [4, Chapter 5] and [1, Chapter 6]. Further details about the performance analysis of digital transmission over fading channels, including the MGF-based method, can be found in [5].

Problems

1. The following bit sequence 0100111011011001 is being transmitted using 16-APSK with constellation mapping given in Figure 4.12. Determine and sketch the modulated symbols.

2. Determine the average symbol energy of 16-APSK modulation scheme with constellation shown in Figure 4.12. The result should be a function of A_1, A_2, and T_s.

3. Consider a system operating over AWGN channel, where the additive white Gaussian noise at the receiver has a power spectrum density of 10^{-4} mW/Hz. The transmission power is 30 mW and the bandwidth of the transmitted signal is 20 kHz. Assuming the system uses a pulse-shaping function that leads to $B_s T_s = 1.2$, determine the exact and approximate symbol error probability of the system with (i) QPSK and (ii) 16-QAM modulation schemes.

4. Determine the phases of modulated symbols for bit stream 0010101100 with $\pi/4$ differential QPSK assuming initial symbol phase equal to zero and the following bit mapping scheme.

Bit pairs	11	01	00	10
Phase difference	$\pi/4$	$3\pi/4$	$-3\pi/4$	$-\pi/4$

If the phase of the channel gain z is equal to $2\pi/3$ during the transmission of these symbols, what are the phase of the received symbols? Can the receiver perfectly recover the transmitted bits if the noise is negligible?

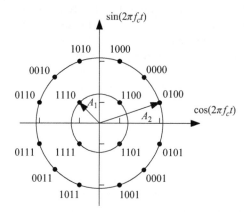

Figure 4.12   Sample constellation mapping for 16-ary amplitude-phase shift keying (16-APSK) modulation

5. Consider the communication over a wireless channel where the transmit power is 10 dBm and the signal bandwidth is 40 kHz. The noise power spectrum density at the receiver $(N_0/2)$ is 10^{-8} mW/Hz.

   (i) Determine the instantaneous bit error rate (BER) of BPSK modulation scheme over this channel when the channel power gain at a particular time instant is -30 dB.

   (ii) Repeat your calculation for the case that the channel power gain becomes -50 dB.

6. Assume a wireless transmission system with binary DPSK modulation scheme is operating over a special fading channel, where the received signal-to-noise ratio (SNR) is equal to 15 dB for the 98% of the time and equal to 3 dB for the 2% of the time.

   (i) Determine the average bit error rate of the wireless system. Note that the instantaneous error rate of DPSK is

   $$P_b(\gamma) = \frac{1}{2}\exp(-\gamma), \tag{4.91}$$

   where γ is the instantaneous SNR.

   (ii) Repeat your calculation for (i) for the following two improved channels: (a) SNR is equal to 20 dB for the 98% of the time and equal to 3 dB for the 2% of the time; (b) SNR is equal to 15 dB for the 98% of the time and equal to 6 dB for the 2% of the time. What can you conclude when comparing the results with part (i)?

7. Assume a Rayleigh fading channel with average channel power gain $2\sigma^2 = -80$ dB. What is the power outage probability of this channel relative to the received signal power threshold $P_0 = -95$ dBm, assuming transmit power P_t is 0 dBm? How about $P_0 = -90$ dBm?

8. Let consider a Rayleigh fading channel with average channel power gain $2\sigma^2 = -60$ dB. The transmit power is 10 dBm, and the signal bandwidth is 40 kHz. The noise power spectrum density at the receiver $(N_0/2)$ is 10^{-5} mW/Hz. What is the outage probability of this channel relative to the SNR threshold of 5 dB? If we want to maintain an outage probability of 0.01 over such channel, what is the smallest transmit power that we can use?

9. The amplitude of the complex channel gain for Rayleigh fading scenario $|z|$ has the following PDF:

   $$p_{|z|}(x) = \frac{x}{\sigma^2}\exp\left[-\frac{x^2}{2\sigma^2}\right], \tag{4.92}$$

   Derive the corresponding PDF of the amplitude square $|z|^2$.

10. Assume that z_I and z_Q are independent Gaussian random variables with zero mean and variance σ^2. Show that the PDF of $|z| = \sqrt{z_I^2 + z_Q^2}$ is

   $$p_{|z|}(x) = \frac{x}{\sigma^2}\exp\left[-\frac{x^2}{2\sigma^2}\right], \tag{4.93}$$

   which implies that $|z|$ is a Rayleigh random variable.

11. The bit error probability of binary frequency shift keying (BFSK) modulation scheme over AWGN channel is

$$P_b(E) = Q(\sqrt{\gamma_s}),\tag{4.94}$$

where γ_s is the channel SNR. Derive a closed-form expression for the average error rate of BFSK over Rayleigh fading channel.

12. Apply the MGF-based approach to analyze the average error rate of BFSK modulation scheme over Rician fading channel.

Bibliography

[1]   J. G. Proakis, *Digital Communications*, 4th ed. New York, NY: McGraw-Hill, 2001.

[2]   T. S. Rappaport, *Wireless Communications: Principle and Practice*, 2nd ed. Upper Saddle River, NJ: Prentice Hall, 2002.

[3]   G. L. Stüber, *Principles of Mobile Communications*, 2nd ed. Norwell, MA: Kluwer Academic Publishers, 2000.

[4]   A. Goldsmith, *Wireless Communications*, New York, NY: Cambridge University Press, 2005.

[5]   M. K. Simon, and M.-S. Alouini, *Digital Communication over Fading Channels*, 2nd ed. New York, NY: John Wiley & Sons, 2005.

Chapter 5
Fading mitigation through diversity combining

Diversity combining is one of the most effective fading mitigation techniques. The basic design principle of diversity technique influences the development of several advanced wireless transmission technology. In this chapter, we study the design and analysis of fundamental diversity combining schemes. After discussing the basic implementation strategies, we investigate several conventional diversity combining schemes, including selection combining, maximum ratio combining, and thresholding combining. Special emphasis is put on the tradeoff analysis of performance versus complexity among different combining schemes. Finally, we presents the transmit diversity solutions for the multiple antennas at the transmitter scenario.

5.1 Basics of diversity combining

The performance of digital transmission schemes degrades dramatically when operating over fading channels. The occurrence of deep fade, although not very frequent, will lead to very small instantaneous received signal-to-noise ratio (SNR), which greatly deteriorates the average error rate performance. Over the past several decades, various technologies have been developed to improve the performance of digital transmission over fading channels. Diversity combining technique is widely recognized as one of the most effectively solution. The basic idea of diversity combining is to somehow create multiple channels to transmit/receive the same information bearing signal. The probability that these channels simultaneously experience deep fade is usually small, especially when they undergo independent fading. By properly combining the received replicas over these channels together, we can improve the quality of received signal and achieve better detection performance.

5.1.1 Implementation strategies

The first step of implementing diversity technique is to create multiple fading channels between transmitter and receiver. Ideally, these channels should experience independent fading process for maximum benefit. An intuitive approach is to use multiple frequency bands to create the channels, which leads to the so-called frequency diversity. If the frequency bands are separated by the order of channel coherence bandwidth B_c, then the signal transmitted over these bands may experience independent fading.

On the other hand, sending the same information signal over different frequency bands will lead to an inefficient utilization of the precious spectrum resource. Another approach is to use different time slots to create multiple channels. If the time slots are separated by the order of channel coherence time T_c, then the transmitted signal over different time slots may experience independent fading. Meanwhile, this time diversity approach will again lead to redundant transmission and, as such, inefficient spectrum utilization. Nevertheless, time and frequency diversity solutions are explored to achieve certain desirable tradeoff between spectral efficiency and reliability, as in multicarrier transmission systems and coding-with-interleaving transmission systems.

Other approaches to implement independent channels include polarization diversity, path diversity, and antenna diversity. Polarization diversity creates two independent channels by using two available polarizations. Path diversity is applicable to wideband transmission systems where the signal received from different propagation paths can be resolved. Antenna diversity uses multiple transmit and receive antennas to create independent channels. In general, these approaches improve the reliability of the system at the cost of higher transceiver complexity. Considering the increasingly scarcity of spectrum resource, these diversity solutions become more attractive, as they provide diversity benefit without sacrificing the spectral efficiency. Among them, antenna diversity solution is widely recognized as one of the most important solution. We will base our following discussion mainly on antenna reception diversity implementation.

5.1.2 Antenna reception diversity

The generic structure of antenna diversity receiver is shown in Figure 5.1. Specifically, the transmitter is equipped with a single antenna whereas the receiver has L antennas. Each receive antenna will collect a copy of the transmitted signal. As such, L parallel channels are created between the transmitter and the receiver at the cost of higher receiver complexity. The received signal copies will be processed by the diversity combiner. The objective of the diversity combiner is to use these L receive signal copies to generate a signal with better quality. Note that the detection will be performed on the combiner output signal. Such antenna reception diversity solution is particularly suitable for uplink transmission, where the base station or access point can afford

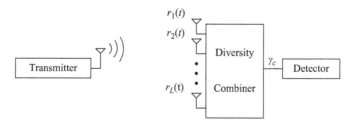

Figure 5.1   Antenna reception diversity system

higher complexity. Meanwhile, the generic diversity receiver structure in Figure 5.1 also applies to other diversity solutions.

In general, diversity technique offers the largest potential performance gain when the signal copies received over different channel experience independent fading. For antenna reception diversity, the complex channel gains corresponding to different receive antennas should be independent of each other. We can create independent fading channels by separating the antennas sufficiently far apart from each other. On the basis of the fading channel model discussed in the previous chapters, the minimum spacing between antennas should be more than half of a wavelength to realize independent fading channels. For example, the wavelength of radio transmission with carrier frequency of 2 GHz will be equal to 0.15 m. As such, the minimum spacing between different receiving antennas should be 75 mm, which can be very difficult to satisfy at the hand-hold devices. The performance degradation due to correlated fading has been investigated in the literature but omitted here for clarity.

For the given diversity receiver structure, the next step in receiver design is to determine appropriate combining scheme. The most popular traditional combining schemes include selection combining, maximal ratio combining, threshold combining, and equal gain combining. In the following section, we will explain the basic ideas of these combining schemes. Naturally, different combining schemes will lead to different tradeoff between performance versus complexity. More specifically, each scheme requires different amount of channel state information to operate and entails different level of hardware complexity while leading to different amount of performance improvement. To quantify such tradeoff, we need to accurately evaluate the performance of different diversity combining schemes.

5.1.3 Framework for performance analysis

We now present a framework for the performance analysis of antenna reception diversity systems, based on which we will analyze and compare the performance of different combining schemes.

We consider a slow frequency-flat fading environment. The fading channel from the transmit antenna to the ith receive antenna is characterized by the complex channel gain $z_i = a_i e^{j\theta_i}, i = 1, 2, \ldots, L$. The received signal on each antenna will be contaminated by additive white Gaussian noise (AWGN) with common power spectral density N_0. Therefore, the SNR at the ith receive antenna, given by $\gamma_i = a_i^2 E_s / N_0$, serves as a suitable channel quality indicators. Due to the fading effect, γ_i will be fluctuating dramatically over time. For the sake of clarity, we assume in the following that the diversity paths experience independent fading. Specifically, the complex channel gain of the ith diversity path, z_i, varies independently over time. We further assume that the antenna spacing is negligible compared to the propagation distance and different antenna experience identical path loss/shadowing effect. It follows that the fading channel power gain a_i^2 for all diversity paths share a common average. Therefore, the received SNR corresponding to the ith diversity path $\gamma_i, i = 1, 2, \ldots, L$, can be modeled as independent and identically distributed (i.i.d.) random variables.

The performance of diversity system will depend on the statistics of the combiner output SNR, denoted by γ_c. In particular, the outage probability, which now becomes the probability that γ_c is smaller than a threshold γ_{th}, can be calculated as

$$P_{out} = \Pr[\gamma_c < \gamma_{th}] = F_{\gamma_c}(\gamma_{th}), \tag{5.1}$$

where $F_{\gamma_c}(\cdot)$ denotes the cumulative distribution function (CDF) of the combined SNR γ_c. The average error rate performance of diversity systems can be evaluated, by averaging the instantaneous bit error rate (BER) over the distribution of γ_c, as

$$\overline{P}_E = \int_0^\infty P_E(\gamma)p_{\gamma_c}(\gamma)d\gamma, \tag{5.2}$$

where $p_{\gamma_c}(\cdot)$ denotes the probability density function (PDF) of the combined SNR γ_c. As such, the performance analysis of diversity systems mandates the statistics of the combiner output SNR. In the following section, we will illustrate the derivation of the CDF and PDF of γ_c from the statistics of the individual path SNR according to the mode of operation of each combining scheme.

5.2   Selection combining

Selection combining (SC) is one of the most popular low-complexity combining scheme. With SC, the diversity branch with the highest instantaneous received SNR is selected and used for data reception. The mode of operation of SC is illustrated in Figure 5.2 for three diversity branch case. Specifically, the diversity combiner will estimate the received SNRs (or equivalently, the received signal power when noise power across branches are identical) of all available diversity branches and compare them. Only the diversity branch currently enjoying the highest received SNR will be used for reception. Such SNR estimation and comparison will be repeated to ensure that the SNR of currently selected branch is always the largest. As a result, the selected branch may vary as the received SNRs on different branches change over

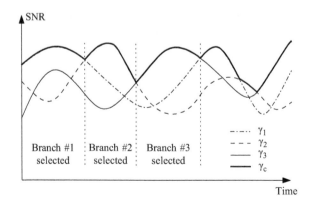

Figure 5.2   Mode of operation of selection combining

time, as shown in Figure 5.2. The instantaneous combiner output SNR with SC is then mathematically given by $\gamma_c = \max\{\gamma_1, \gamma_2, \ldots, \gamma_L\}$, where γ_is are the instantaneous received SNR of the ith diversity branch. Therefore, the combiner output SNR is the largest one of L independent random variables.

We now derive the statistics of γ_c on the basis of the mode of operation of SC. Specifically, the CDF of the combined SNR γ_c can be calculated as

$$F_{\gamma_c}(\gamma) = \Pr[\gamma_c < \gamma] = \Pr[\max\{\gamma_1, \gamma_2, \ldots, \gamma_L\} < \gamma], \tag{5.3}$$

which is the same as the joint probability of $\gamma_i < \gamma$, $i = 1, 2, \ldots, L$. Under the assumption of independent fading, we can rewrite the CDF of γ_c as

$$F_{\gamma_c}(\gamma) = \prod_{i=1}^{L} \Pr[\gamma_i < \gamma] = \prod_{i=1}^{L} F_{\gamma_i}(\gamma), \tag{5.4}$$

where $F_{\gamma_i}(\cdot)$ denotes the CDF of γ_i. If γ_i can be further assumed to be identically distributed with common CDF $F_\gamma(\cdot)$, the CDF of γ_c with SC can be simply written as

$$F_{\gamma_c}(\gamma) = [F_\gamma(\gamma)]^L. \tag{5.5}$$

Finally, the PDF of γ_c with i.i.d. fading on different diversity branches can be obtained, by taking derivative of $F_{\gamma_c}(\gamma)$ with respect to γ, as

$$p_{\gamma_c}(\gamma) = L[F_\gamma(\gamma)]^{L-1} p_\gamma(\gamma), \tag{5.6}$$

where $p_\gamma(\cdot)$ denotes the common PDF of the branch SNR γ_is. These generic distribution functions apply to different fading environment after substituting proper distribution functions of branch SNRs in Table 4.1. Specifically, for i.i.d. Rayleigh fading environment, the PDF of γ_c specializes to

$$p_{\gamma_c}(\gamma) = \frac{L}{\bar{\gamma}}\left[1 - \exp\left(-\frac{\gamma}{\bar{\gamma}}\right)\right]^{L-1} \exp\left(-\frac{\gamma}{\bar{\gamma}}\right), \tag{5.7}$$

where $\bar{\gamma}$ denotes the common average branch SNR. We can now analyze the performance of wireless transmission system with SC diversity receiver. The outage probability can be calculated by evaluating the CDF of γ_c at the outage threshold γ_{th} as

$$P_{out} = [F_\gamma(\gamma_{th})]^L. \tag{5.8}$$

Example: Outage probability of SC

Consider a digital wireless transmission system using BPSK modulation scheme. To satisfy the instantaneous BER requirement of 10^{-3}, the received SNR should be at least 7 dB for BPSK. An L-branch selection combiner is implemented at the receiver to combat the fading effect. The channel introduces i.i.d. Rayleigh fading with average received SNR of 15 dB on each antenna branch. Determine the outage probability for $L = 1, 2$, and 3 cases.

Solutions: We first note that for i.i.d. Rayleigh fading environment, the SNR of individual path, γ_i, shares a common CDF given by

$$F_\gamma(\gamma) = 1 - \exp\left(-\frac{\gamma}{\overline{\gamma}}\right),$$

(5.9)

where the average SNR $\overline{\gamma} = 15\,\text{dB} = 31.6$. When $L = 1$, which corresponds to the no diversity case, the outage probability can be calculated, while noting $\gamma_{\text{th}} = 7\,\text{dB} = 5$, as

$$P_{\text{out}} = F_\gamma(5) = 1 - \exp\left(-\frac{5}{31.6}\right) = 0.1467.$$

(5.10)

When $L = 2$, the outage probability becomes

$$P_{\text{out}} = [F_\gamma(5)]^2 = \left[1 - \exp\left(-\frac{5}{31.6}\right)\right]^2 = 0.0215.$$

(5.11)

and for $L = 3$, the outage probability can be similarly determined to be 0.0031. Apparently, the outage probability reduces approximately by an order of magnitude with the each additional antenna. Diversity technique can effectively improve the outage performance of wireless transmission.

The average error rate of SC can be calculated with the application of the distribution function of the combined SNR γ_c into (5.2).

Example: Average error rate of SC

Determine a closed-form expression for the average bit error rate of binary DPSK transmission over i.i.d. Rayleigh fading with L-branch SC receiver.

Solutions: The average BER of binary DPSK modulation over i.i.d. Rayleigh fading with L-branch SC receiver can be calculated as

$$\overline{P}_E = \int_0^\infty P_E(\gamma) p_{\gamma_c}(\gamma) d\gamma$$

$$= \int_0^\infty \frac{1}{2} \exp(-\gamma) \frac{L}{\overline{\gamma}} \left[1 - \exp\left(-\frac{\gamma}{\overline{\gamma}}\right)\right]^{L-1} \exp\left(-\frac{\gamma}{\overline{\gamma}}\right) d\gamma.$$

(5.12)

After carrying out integration and some manipulation, we arrive at the following closed-form expression for the average BER given by

$$\overline{P}_E = \frac{L}{2} \sum_{l=0}^{L-1} \binom{L-1}{l} (-1)^l (1 + l + \overline{\gamma})^{-1}.$$

(5.13)

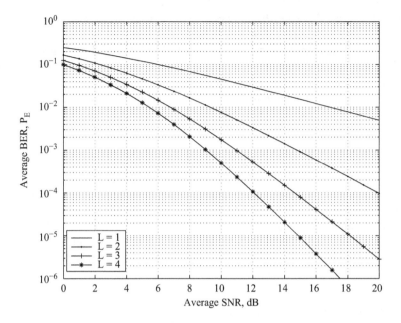

Figure 5.3   Average BER of binary DPSK with selection combining

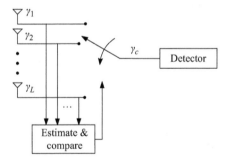

Figure 5.4   Structure of diversity receiver with selection combining

In Figure 5.3, we plot the average BER of binary DPSK with SC receiver as function of the average received SNR per branch $\overline{\gamma}$ for different number of diversity branches. The error rate performance improves as the number of diversity branches increases, but with diminishing gain. SC can effectively improve the average error rate performance of digital transmission systems over fading channel.

The major advantage of SC scheme is the relatively low receiver complexity. The receiver structure with SC is shown in Figure 5.4. Note that the receiver only needs to process the signal from one diversity branch after proper selection. As such, a single RF chain with a switching mechanism is required. In fact, SC diversity offers the best performance when the receiver can only afford a single RF chain. Meanwhile, the receiver will estimate the channel quality on different diversity branch and compare them to determine the best antenna branch, i.e., the one enjoying the highest received SNR. As such, SC diversity requires the quality estimation of all available diversity branches, to ensure that the best branch is always chosen.

5.3   Maximum ratio combining

The best selection strategy with SC diversity neglects the contribution from other diversity branches. Intuitively, a better performance should be achievable if we can exploit the signal replicas from all available diversity branches. In this scenario, maximum ratio combining (MRC) is the most popular linear combining scheme, as it achieves the best performance in the noise limited environment among all linear schemes.

5.3.1   Optimal linear combining

The general structure of a linear combiner is shown in Figure 5.5. Specifically, the combined signal $r_c(t)$ is generated as the linear combination of the received signal replicas on L diversity branches, i.e.,

$$r_c(t) = w_1 \cdot r_1(t) + w_2 \cdot r_2(t) + \cdots + w_L \cdot r_L(t) = \sum_{i=1}^{L} w_i r_i(t), \tag{5.14}$$

where w_i, $i = 1, 2, \ldots, L$, are the combining weights to be determined. The basic design principle of the MRC scheme is to optimally choose w_is such that the SNR of the combined signal $r_c(t)$ is maximized.

For the flat fading environment under consideration, the received signal on the ith diversity branch is given by

$$r_i(t) = a_i e^{j\theta_i} s(t) + n_i(t), \tag{5.15}$$

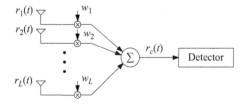

Figure 5.5   General structure of linear combiner

where $s(t)$ is the transmitted signal, $a_i e^{j\theta_i}$ is the complex channel gain, and $n_i(t)$ is the additive white Gaussian noise for the i diversity branch. After proper substitution, the combined signal can be written as

$$r_c(t) = \sum_{i=1}^{L} w_i a_i e^{j\theta_i} s(t) + \sum_{i=1}^{L} w_i n_i(t). \tag{5.16}$$

The SNR of $r_c(t)$ can be shown to be given, while assuming the noise signals are independent with identical noise variance N_0, by

$$\gamma_c = \frac{\left| \sum_{i=1}^{L} w_i \cdot a_i e^{j\theta_i} \right|^2 E_s}{\sum_{i=1}^{L} |w_i|^2 N_0}. \tag{5.17}$$

To find the optimal weights w_is that maximize γ_c, we invoke the Cauchy–Schwartz inequality, which states that

$$\left| \sum_{i=1}^{L} a_i \cdot b_i \right|^2 \leq \left(\sum_{i=1}^{L} |a_i|^2 \right) \left(\sum_{i=1}^{L} |b_i|^2 \right), \tag{5.18}$$

where equality holds when a_i is proportional to the complex conjugate of b_i, i.e., $a_i \propto b_i^*$ for all i. It follows that

$$\gamma_c \leq \frac{\left(\sum_{i=1}^{L} |w_i|^2 \right) \cdot \left(\sum_{i=1}^{L} |a_i e^{j\theta_i}|^2 \right) E_s}{\sum_{i=1}^{L} |w_i|^2 N_0} = \sum_{i=1}^{L} a_i^2 \frac{E_s}{N_0} = \sum_{i=1}^{L} \gamma_i. \tag{5.19}$$

From the above analysis, we conclude that (i) the optimal weight for ith diversity branch w_i should be proportional to the complex conjugate of the corresponding channel gain, i.e., $w_i \propto a_i e^{-j\theta_i}$; (ii) the maximal SNR of combined signal with linear combining is equal to $\gamma_c = \sum_{i=1}^{L} \gamma_i$. The resulting scheme is termed as maximum ratio combining because it maximizes the SNR of combined signal. Intuitively, MRC receiver should achieve better performance than SC receiver, as it explores the contribution of all diversity branches. Note that the combined SNR with SC is $\max\{\gamma_1, \gamma_2, \ldots, \gamma_L\}$, which is definitely smaller than that with MRC, given by $\sum_{i=1}^{L} \gamma_i$.

Meanwhile, MRC receiver has much higher complexity than SC receiver. The structure of an MRC receiver is shown in Figure 5.6. Specifically, the receiver first

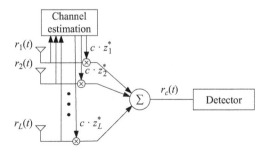

Figure 5.6   Structure of an MRC-based diversity combiner

estimates the complex channel gain of each diversity branch and determines the optimal weights $w_i = c \cdot z_i^* = c a_i e^{-j\theta_i}$, $i = 1, 2, \ldots, L$, where c is a normalizing constant. Then the received signal from each branch is multiplied by the corresponding weight before being summed together, to generate the combiner output signal. Note that the MRC receiver processes the received signal from all diversity branches in a coherent fashion. In addition, the MRC combining operation requires the complete channel knowledge (both amplitude and phase) of all diversity branches. Meanwhile, SC combiner only needs to know the relative strength of different diversity branches and processes the selected branch without requiring the channel phase information.

5.3.2 *Performance analysis*

To quantify the performance advantage of MRC over SC scheme, we analyze the performance of digital wireless transmission with MRC receiver. For that purpose, we need the statistics of the combined SNR with MRC, which is the sum of L random variables, i.e., $\gamma_c = \sum_{i=1}^{L} \gamma_i$. Determining the distribution functions of the sum of random variables is in general a challenging problem even in the statistics literature. Starting from the joint distribution of the individual random variables, the PDF of the sum can be obtained by carrying out $L - 1$-fold of integrations, i.e.,

$$
p_{\gamma_c}(\gamma) = \int_0^\gamma \int_0^{\gamma - \gamma_{L-1}} \cdots \int_0^{\gamma - \sum_{i=2}^{L-1} \gamma_i} p_{\gamma_1, \gamma_2, \ldots, \gamma_L}\left(\gamma_1, \gamma_2, \ldots, \gamma_{L-1}, \gamma - \sum_{i=1}^{L-1} \gamma_i\right)
$$
$$
\times d\gamma_1 d\gamma_2 \cdots d\gamma_{L-1}, \tag{5.20}
$$

where $p_{\gamma_1, \gamma_2, \ldots, \gamma_L}(\ldots)$ is the joint PDF of $\gamma_1, \gamma_2, \ldots$, and γ_L. When the random variables are independent, the PDF of the sum becomes the convolution of L individual PDFs, which is still difficult to calculate directly.

The MGF-based approach becomes an attractive alternative for the analysis of MRC scheme when diversity branches experience independent fading, and as such, branch SNR γ_is are independent random variables. Note that the MGF of the sum of independent random variables is the product of the MGFs of individual random variables. As such, the MGF of the combined SNR with MRC γ_c over independent fading paths can be written as

$$
\mathcal{M}_{\gamma_c}(s) = \prod_{i=1}^{L} \mathcal{M}_{\gamma_i}(s), \tag{5.21}
$$

where $\mathcal{M}_{\gamma_i}(s)$ is the MGF of the received SNR over the ith diversity branch. After substituting the MGF of individual branch SNRs (see Table 4.1), we can readily obtain $\mathcal{M}_{\gamma_c}(s)$ for most fading channel models of interest. For example, the MGF of combined SNR with MRC over i.i.d. Rayleigh fading channels is given by

$$
\mathcal{M}_{\gamma_c}(s) = (1 - s\bar{\gamma})^{-L}. \tag{5.22}
$$

Starting from the MGF, we can determine the PDF of γ_c with MRC after applying proper inverse Laplace transform. For i.i.d. Rayleigh fading scenario, the PDF of the combined SNR can be obtained as

$$p_{\gamma_c}(\gamma) = \frac{\gamma^{L-1} e^{-\gamma/\overline{\gamma}}}{\overline{\gamma}^L (L-1)!}. \tag{5.23}$$

The CDF of γ_c can be routinely obtained.

With the statistics of the combined SNR derived, we can analyze the performance of MRC scheme. For example, the outage probability of digital transmission with MRC receiver in i.i.d. Rayleigh fading environment can be determined as

$$P_{\text{out}} = \int_0^{\gamma_{\text{th}}} p_{\gamma_c}(\gamma) d\gamma \tag{5.24}$$

$$= 1 - \exp(-\gamma_{\text{th}}/\overline{\gamma}) \sum_{l=0}^{L-1} \frac{(\gamma_{\text{th}}/\overline{\gamma})^l}{l!}.$$

Example: Outage probability of MRC

A digital wireless transmission system with L-branch MRC receiver is operating over i.i.d. Rayleigh fading. The average received SNR on each diversity branch is 15 dB. Determine the outage probability for $L = 1, 2,$ and 3 cases when the outage threshold is 7 dB.

Solutions: To solve this problem, we can directly apply the result of (5.24) with $\overline{\gamma} = 15\,\text{dB} = 31.6$ and $\gamma_{\text{th}} = 7\,\text{dB} = 5$. When $L = 1$, which corresponds to the no diversity case, (5.24) specializes to

$$P_{\text{out}} = 1 - \exp\left(-\frac{5}{31.6}\right) = 0.1467. \tag{5.25}$$

When $L = 2$, (5.24) becomes

$$P_{\text{out}} = 1 - \exp\left(-\frac{5}{31.6}\right)\left(1 + \frac{5}{31.6}\right) = 0.0116. \tag{5.26}$$

and for $L = 3$, we have

$$P_{\text{out}} = 1 - \exp\left(-\frac{5}{31.6}\right)\left(1 + \left(\frac{5}{31.6}\right) + \frac{1}{2}\left(\frac{5}{31.6}\right)^2\right) = 0.00091. \tag{5.27}$$

Compared with the results in previous example on the outage probability of SC, we can see that MRC leads to smaller outage probability under the same parameter setting. The performance advantage of MRC comes at the cost of higher receiver complexity.

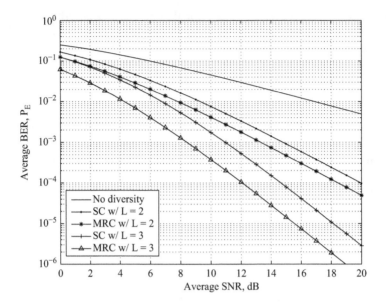

Figure 5.7   Average error rate comparison between SC and MRC

The average error rate performance can be analyzed using either the PDF or the MGF approach. Specifically, the average bit error rate of BPSK modulation scheme over i.i.d. Rayleigh fading can be calculated, after applying the PDF of γ_c given in (5.23), as

$$
\begin{aligned}
\overline{P}_E &= \int_0^\infty Q\left(\sqrt{2\gamma}\right) \frac{\gamma^{L-1} e^{-\gamma/\overline{\gamma}}}{\overline{\gamma}^L (L-1)!} d\gamma \\
&= \left(\frac{1 - \sqrt{\overline{\gamma}/(1+\overline{\gamma})}}{2}\right)^L \sum_{l=0}^{L-1} \binom{L-1+l}{l} \left(\frac{1 - \sqrt{\overline{\gamma}/(1+\overline{\gamma})}}{2}\right)^l.
\end{aligned}
\tag{5.28}
$$

The average bit error rate of binary DPSK modulation scheme, whose instantaneous bit error probability is $P_b = e^{-\gamma}/2$, over i.i.d. Rayleigh fading can be evaluated as

$$
\overline{P}_E = \frac{1}{2} \mathcal{M}_{\gamma_c}(-1) = \frac{1}{2}(1 + \overline{\gamma})^{-L}.
\tag{5.29}
$$

Figure 5.7 compares the average error rate of binary DPSK with SC and MRC schemes. While both diversity schemes achieve considerable performance gain over no diversity scenario, MRC achieves much better average error rate performance than SC with the same number of diversity branches. The performance gaps between MRC and SC grows as the number of diversity branches increases. This can be explained by the fact that MRC exploits the contribution of all diversity branches whereas SC only relies on the best branch. The benefit of combining all branches grows faster than selecting the best when the number of branches increases, but at the cost of increasing receiver complexity. As an intermediate solution of these two extremes, combining several best branches in the MRC fashion will lead to a better performance versus complexity tradeoff when the number of available diversity branches is very large.

5.3.3 Equal gain combining

Equal gain combining (EGC) is another combining scheme that exploits the contribution of all diversity branches, but with slightly lower complexity than MRC. With EGC, the combined signal is still the linear combination of the signals received from different branches. The weight for the ith diversity branch is $w_i = e^{-j\theta_i}$, which is simpler than those for MRC. The combined signal with EGC is then given by

$$
r_c(t) = \sum_{i=1}^{L} e^{-j\theta_i} r_i(t)
$$

$$
= \sum_{i=1}^{L} a_i s(t) + \sum_{i=1}^{L} e^{-j\theta_i} n_i(t). \tag{5.30}
$$

The SNR of the combined signal can be shown to be equal to

$$
\gamma_c = \frac{E_s}{N_0 L} \left(\sum_{i=1}^{L} \alpha_i \right)^2, \tag{5.31}
$$

where E_s is the symbol energy and N_0 is the noise spectrum density. Essentially, the EGC combiner cophases the signals received on different diversity branches and adds them together with unit weight. As such, the receiver with EGC only needs to estimate the channel phase of each diversity path. In general, it is more challenging to estimate the channel phase than the channel amplitude. Furthermore, SC needs only to process the currently selected branch, whereas EGC receiver still needs to process all L available branches. Therefore, EGC entails higher implementation complexity than SC.

Example: Combined SNR with SC, MRC, and EGC

A digital wireless transmission system is operating over frequency flat fading environment. The SNR per symbol at the transmitter E_s/N_0 is 10 dB. A three-branch diversity receiver is deployed. The instantaneous complex channel gains of three diversity branches are $\alpha_1 e^{j\theta_1} = 0.2e^{j\pi/3}$, $\alpha_2 e^{j\theta_2} = 0.35e^{j3\pi/5}$, and $\alpha_3 e^{j\theta_3} = 0.12e^{j7\pi/4}$. Determine the instantaneous SNR of the combined signal when the receiver employs (i) SC, (ii) MRC, and (iii) EGC schemes.

Solutions:

(i)   With SC, the branch that leads to the largest instantaneous SNR is selected. Since the second branch has the largest instantaneous power gain of $\alpha_2^2 = 0.35^2 = 0.1225$, it will be used, which results the instantaneous combined SNR of 1.225, or equivalently, 0.88 dB.

(ii) With MRC, the combined SNR is the sum of branch SNRs, which can be calculated as

$$\gamma_c = \gamma_1 + \gamma_2 + \gamma_3 = (0.2^2 + 0.35^2 + 0.12^2)E_s/N_0 = 1.769 = 2.48\,\text{dB}.$$

$$(5.32)$$

(iii) With EGC, the instantaneous combined SNR can be determined using (5.31) as

$$\gamma_c = [(0.2 + 0.35 + 0.12)^2/3]E_s/N_0 = 1.496, \qquad (5.33)$$

which is equal to 1.75 dB.

5.4   Threshold combining and its variants

Threshold combining is another classical combining scheme with even lower complexity than SC. Note that the receiver with SC needs to monitor and compare the quality of all available diversity branches in order to determine the best branch. The receiver with threshold combining only monitors the quality of the currently used branch. If the quality of the current branch is unacceptable, the receiver will switch to another available branch. As such, this type of combining schemes are also known as switched combining schemes. The structure of diversity receiver with threshold combining is illustrated in Figure 5.8. Note that the channel estimation is performed only on the selected branch.

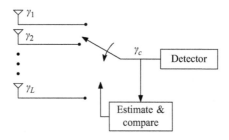

Figure 5.8   Structure of diversity receiver with threshold combining

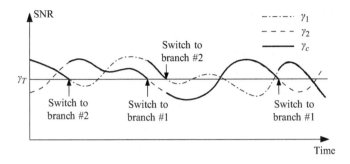

Figure 5.9   Mode of operation of dual branch switch and stay combining

5.4.1   Dual-branch switch and stay combining

Dual-branch switch and stay combining (SSC) is the most well-known sample threshold combining scheme. With SSC, the receiver estimates the SNR of the currently used branch (or received signal power) and compares it with a fixed threshold, denoted by γ_T. If the estimated SNR is greater or equal to γ_T, then the current branch is deemed acceptable and the receiver continues to use it for data reception. Otherwise, the receiver will switch and use the other branch for data reception, regardless of its quality. The mode of operation of SSC is illustrated in Figure 5.9. We can see from the figure that the receiver with SSC scheme may switch from an unacceptable branch to an even worse branch and use it for reception.

To analyze the performance of SSC scheme, we need to derive the statistics of the combined SNR with SSC. Note that the operation of SSC can be mathematically summarized as

$$\gamma_c = \begin{cases} \gamma_1, & \gamma_1 \geq \gamma_T; \\ \gamma_2, & \gamma_1 < \gamma_T. \end{cases} \tag{5.34}$$

Here, we assume that γ_1 is the SNR of the currently used branch, without loss of generality. Applying the total probability theorem, we can calculate the CDF of the combined SNR, γ_c, as

$$F_{\gamma_c}(\gamma) = \Pr\left[\gamma_1 < \gamma, \gamma_1 \geq \gamma_T\right] + \Pr\left[\gamma_2 < \gamma, \gamma_1 < \gamma_T\right]. \tag{5.35}$$

Assuming i.i.d. fading branches, the CDF of γ_c can be obtained, in terms of the common CDF of branch SNR, $F_\gamma(\gamma)$, as

$$F_{\gamma_c}(\gamma) = \begin{cases} F_\gamma(\gamma_T)F_\gamma(\gamma) + F_\gamma(\gamma) - F_\gamma(\gamma_T), & \gamma \geq \gamma_T; \\ F_\gamma(\gamma_T)F_\gamma(\gamma), & \gamma < \gamma_T. \end{cases} \tag{5.36}$$

It follows, after taking derivative with respect to γ, that the PDF of the combined SNR with SSC is given by

$$p_{\gamma_c}(\gamma) = \begin{cases} F_\gamma(\gamma_T)p_\gamma(\gamma) + p_\gamma(\gamma), & \gamma \geq \gamma_T; \\ F_\gamma(\gamma_T)p_\gamma(\gamma), & \gamma < \gamma_T, \end{cases} \tag{5.37}$$

where $p_\gamma(\gamma)$ denotes the common PDF of branch SNR. The statistics of the combined SNR for more general scenario with path correlation and/or unbalanced paths can also be obtained by applying a Markov-chain-based approach.

We can apply these generic results on the statistics of combined SNR to analyze the performance of SSC over various fading environment. For example, the out-age probability of digital transmission with SSC receiver over i.i.d. Rayleigh fading environment can be calculated as

$$P_{\text{out}} = \begin{cases} 1 - 2\exp(-\gamma_{\text{th}}/\overline{\gamma}) + \exp(-(\gamma_T + \gamma_{\text{th}})/\overline{\gamma}), & \gamma_{\text{th}} \geq \gamma_T; \\ (1 - \exp(-\gamma_{\text{th}}/\overline{\gamma}))(1 - \exp(-\gamma_T/\overline{\gamma})), & \gamma_{\text{th}} < \gamma_T. \end{cases} \tag{5.38}$$

Meanwhile, given a particular modulation scheme with instantaneous error rate of $P_E(\gamma)$, the generic expression of the average error rate with SSC receiver can be calculated as

$$\overline{P}_E = F_\gamma(\gamma_T) \int_0^\infty P_E(\gamma)p_\gamma(\gamma)d\gamma + \int_{\gamma_T}^\infty P_E(\gamma)p_\gamma(\gamma)d\gamma. \tag{5.39}$$

Example: Average error rate of SSC

Determine an closed-form expression for the average bit error rate of binary DPSK transmission over i.i.d. Rayleigh fading with dual-branch SSC receiver.

Solutions: The average BER of binary DPSK modulation over i.i.d. Rayleigh fading with dual-branch SSC receiver can be calculated as

$$\overline{P}_E = \int_0^\infty \frac{1}{2}\exp(-\gamma)p_{\gamma_c}(\gamma)d\gamma$$

$$= [1 - \exp(-\gamma_T/\overline{\gamma})] \int_0^\infty \frac{1}{2}\exp(-\gamma)\frac{1}{\overline{\gamma}}\exp(-\gamma/\overline{\gamma})d\gamma$$

$$+ \int_{\gamma_T}^\infty \frac{1}{2}\exp(-\gamma)\frac{1}{\overline{\gamma}}\exp(-\gamma/\overline{\gamma})d\gamma. \tag{5.40}$$

After carrying out integration and some manipulation, we arrive at the following closed-form expression for the average BER given by

$$\overline{P}_E = \frac{1}{2(1 + \overline{\gamma})}(1 - \exp(-\gamma_T/\overline{\gamma}) + \exp(-\gamma_T)\exp(-\gamma_T/\overline{\gamma})). \tag{5.41}$$

In Figure 5.10, we plot the average BER of binary DPSK with SSC receiver as function of the SNR threshold γ_T. The average BER for no diversity and dual branch SC cases are also plotted. We can observe that there is an optimal value of the SNR

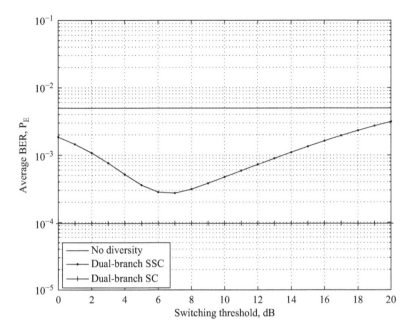

Figure 5.10 Average bit error rate of binary DPSK with SSC receiver ($\overline{\gamma} = 20\,dB$)

threshold γ_T, in terms of minimizing the average BER of SSC receiver. Note that when γ_T is very small, the receiver will always use one branch. Meanwhile, if γ_T is very large, the receiver keeps switching to the other branch. Both cases will lead to the same performance as no diversity scenario.

5.4.2 Multibranch switch and examine combining

SSC scheme will have the same performance even there are more than two branches as the receiver will stay on the switch-to branch. Switch and examine combining (SEC) scheme is another threshold combining scheme that can exploit the diversity benefit of additional diversity branches. The receiver with SEC will examine the quality of the switched-to branch and switch again if it finds the quality of the switch-to branch unacceptable. This process is continued until either an acceptable branch is found or all branches have been examined. The mode of operation of L-branch SEC can be mathematically summarized, assuming the first branch is the currently used branch, as

$$\gamma_c = \begin{cases} \gamma_1, & \gamma_1 \geq \gamma_T; \\ \gamma_2, & \gamma_1 < \gamma_T, \gamma_2 \geq \gamma_T; \\ \gamma_3, & \gamma_1 < \gamma_T, \gamma_2 < \gamma_T, \gamma_3 \geq \gamma_T; \\ \vdots & \vdots \\ \gamma_L, & \gamma_i < \gamma_T, i = 1, 2, \ldots, L-1. \end{cases} \quad (5.42)$$

Following the similar analytical approach in the previous subsection, the CDF of the combined SNR with L-branch SEC over i.i.d. fading branches can be obtained as

$$F_{\gamma_c}(x) = \begin{cases} [F_\gamma(\gamma_T)]^{L-1} F_\gamma(x), & x < \gamma_T; \\ \displaystyle\sum_{j=0}^{L-1} [F_\gamma(x) - F_\gamma(\gamma_T)][F_\gamma(\gamma_T)]^j \\ \quad + [F_\gamma(\gamma_T)]^L, & x \geq \gamma_T. \end{cases} \tag{5.43}$$

The corresponding PDF and MGF of combined SNR γ_c can be then routinely obtained and applied to its performance analysis of SEC over fading channels.

5.4.3 Threshold combining with post-examining selection

Another variant of threshold combining is the so-called switch and examine combining with post-examining selection (SECps). Similar to the SSC/SEC schemes, the receiver with SECps tries to use acceptable diversity branches by examining as many branches as necessary. But when no acceptable path is found after examining all available branches, the receiver with SECps will use the best unacceptable one. The operation of dual branch SECps is illustrated in Figure 5.11. Note that compared with dual SSC scheme, SECps scheme will lead to better output signal when neither diversity branch is acceptable.

The mode of operation of L-branch SECps can be mathematically summarized as

$$\gamma_c = \begin{cases} \gamma_1, & \gamma_1 \geq \gamma_T; \\ \gamma_2, & \gamma_1 < \gamma_T, \gamma_2 \geq \gamma_T; \\ \gamma_3, & \gamma_1 < \gamma_T, \gamma_2 < \gamma_T, \gamma_3 \geq \gamma_T; \\ \vdots & \vdots \\ \max\{\gamma_1, \gamma_2, \ldots, \gamma_L\}, & \gamma_i < \gamma_T, i = 1, 2, \ldots, L - 1. \end{cases} \tag{5.44}$$

Note that when first $L - 1$ branches are unacceptable, the receiver will essentially perform SC on all branches regardless whether the L branch is acceptable or not.

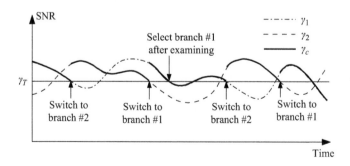

Figure 5.11   Mode of operation of SECps scheme over dual diversity branches

Consequently, the CDF of the combined SNR with SECps over i.i.d. fading branches can be shown to be given by

$$
F_{\gamma_c}(x) = \begin{cases} 1 - \displaystyle\sum_{i=0}^{L-1} [F_\gamma(\gamma_T)]^i [1 - F_\gamma(x)], & x \geq \gamma_T; \\ [F_\gamma(x)]^L, & x < \gamma_T, \end{cases} \tag{5.45}
$$

which can be applied to the outage performance analysis of SECps scheme.

Figure 5.12 shows the outage performance of different threshold combining schemes under consideration in comparison with those of no diversity and SC cases. We first note that threshold combining, including the basic SSC scheme, can effectively improve the outage performance of wireless transmission. The SEC scheme can benefit from the additional diversity branches when the number of diversity branches increases from two to four. In fact, 4-branch SEC has exact the same performance as 4-branch SC when the switching threshold is equal to the outage threshold. It is also interesting to see that 4-branch SECps has the same outage probability as 4-branch SC when $\gamma_{th} \leq \gamma_T$ and as 4-branch SEC when $\gamma_{th} > \gamma_T$. Intuitively, when $\gamma_{th} \leq \gamma_T$, the receiver with SECps experiences outage only if the SNR of all diversity branches falls below γ_T and their maximum is smaller than γ_{th}. On the other hand, when $\gamma_{th} > \gamma_T$, the receiver with SECps experiences outage if the SNR of the first acceptable branch is smaller than γ_{th} or when the SNR of all diversity branches fall below γ_T.

Meanwhile, different variants of threshold combining entail different operational complexity. Specifically, dual-branch SSC needs only one branch estimation for its

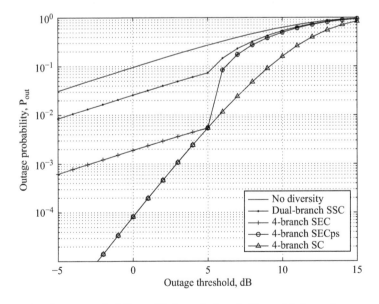

Figure 5.12 Outage probability of L-branch threshold combining schemes as function of outage threshold γ_{th} ($\gamma_T = 5\,dB$, $\overline{\gamma} = 10\,dB$)

operation whereas SEC and SECps typically requires multiple branch estimation. SSC receiver performs at most one branch switching during branch update, where SEC and SECps scheme may perform multiple branch switching. SSC and SEC schemes only needs to compare the branch SNR with a fixed threshold, whereas SECps and SC scheme needs to compare branch SNRs to determine the best. In the worst case scenario, SECps has the same operational complexity as the SC scheme.

5.5   Transmit diversity

So far, we have been considering the transmission scenario where there is a single transmit antenna and multiple receive antennas, which is typically applicable to the uplink transmission from mobile terminal to base station or access point. When there are multiple transmit antennas and a single receive antenna, as is often the case in the downlink transmission, we can apply transmit diversity techniques to explore diversity benefit. The generic structure of diversity transmission system is show in Figure 5.13. The information signal $s(t)$ is processed by diversity transmitter to generate signals $x_i(t)$, $i = 1, 2, \ldots, L$, for transmission on L transmit antennas. Transmit diversity schemes can be classified into two categories, depending on whether the channel state information (CSI) is available at the transmitter side or not. If the transmitter has the necessary knowledge about the channel, then conventional diversity schemes can be readily adapted and applied to the transmit diversity scenario. On the other hand, when the transmitter has no channel knowledge, the system can still explore the diversity benefit by applying certain space-time-coded transmission schemes.

5.5.1   Channel knowledge available at transmitter

If the appropriate CSI corresponding to different transmit antennas are available at the transmitter side, then most conventional combining schemes discussed in the previous sections can apply. For example, if the complex gain $a_i e^{j\theta_i}$ for the channel from the ith transmit antenna to the receive antenna is available at the transmitter, we can implement the so-called transmit MRC, also known as maximum ratio transmission (MRT), by multiplying the information signal $s(t)$ by a weight w_i, which is proportional

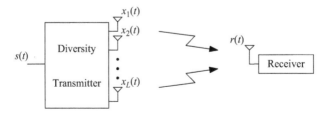

Figure 5.13   Transmit diversity scenario

to $a_i e^{-j\theta_i}$, and then transmitting it on the ith antenna. To satisfy the total transmit power constraint, we can choose the amplitude of w_i as

$$|w_i| = \frac{a_i}{\sqrt{\sum_{j=1}^{L} a_j^2}}, \quad i = 1, 2, \dots, L, \tag{5.46}$$

which leads to $\sum_{i=1}^{L} |w_i|^2 = 1$. Then, the transmitted signal from the ith transmit antenna is given by

$$x_i(t) = \frac{a_i}{\sqrt{\sum_{j=1}^{L} a_j^2}} e^{-j\theta_i} s(t), \quad i = 1, 2, \dots, L. \tag{5.47}$$

After propagating through the fading channel, the signal transmitted from different antennas will add up at the receiver. Specifically, the received signal is given by

$$r(t) = a_1 e^{j\theta_1} w_1 s(t) + a_2 e^{j\theta_2} w_2 s(t) + \dots + a_L e^{j\theta_L} w_l s(t) + n(t)$$

$$= \sqrt{\sum_{j=1}^{L} a_j^2} s(t) + n(t), \tag{5.48}$$

where $n(t)$ is the additive white Gaussian noise. Essentially, the information signal transmitted from different antenna add together coherently after being weighted proportionally to the corresponding channel power gains. As such, the SNR of the received signal is $\gamma_r = \sum_{i=1}^{L} \gamma_i$, which is the same of the combiner output SNR with receive MRC. Note that the total transmission power of transmit MRC is the same as conventional system. Therefore, transmit MRC achieves the same diversity benefit as receiver MRC. Figure 5.14(a) illustrates the structure of the transmit MRC-based diversity transmission system.

SC can also be implemented at the transmitter side, leading to the so-called transmit antenna selection (TAS) scheme. Essentially, the antenna that leads to the highest received SNR will be selected and used for transmission. The received SNR with TAS will be equal to $\gamma_r = \max_i \{\gamma_i\}$. As such, TAS achieves the same diversity benefit as their receive diversity counterpart. The transmitter with TAS only needs to implement an antenna switching structure and switches to the best antenna based on the channel state information, as shown in Figure 5.14(b). Therefore, TAS has much

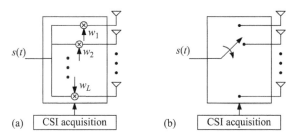

Figure 5.14   *Transmit diversity with channel state information at the transmitter: (a) transmit MRC and (b) transmit antenna selection*

lower complexity than transmit MRC. Threshold combining can also be implemented in the transmit diversity setting, leading to even lower complexity.

The main challenge in implementing these conventional diversity schemes in transmit diversity setting is to acquire the channel knowledge at the transmitter side. For the general two-way transmission scenario under consideration, there are two approaches to make CSI available at the transmitter side. The first approach implies receiver feedback over the reverse channel. As discussed in the earlier sections, the channel knowledge can be conveniently obtained at the receiver side through the channel estimation process. Using the pilot symbols sent by the transmitter, the receiver can estimate the channel gains and send the required channel knowledge back to the transmitter. This approach is particularly applicable to wireless systems adopting frequency division duplexing (FDD) scheme.

Different transmit diversity schemes requires different amount of channel knowledge, which in turn affects the amount of feedback over reverse channel. Specifically, for transmit MRC, the receiver needs to feedback the complex channel gain corresponding to each transmit antenna, whereas for TAS, the receiver just needs to feedback the index of the antenna that leads to the highest received SNR. The implementation of MRC will require the feedback load of L complex numbers, whereas TAS requires a feedback load of $\lceil \log_2 L \rceil$ bits, where L is the number of transmit antenna. From this perspective, TAS is more attractive than transmit MRC in the transmit diversity setting. Finally, the receiver with transmit threshold combining just needs to feedback one bit of information to indicate whether the transmitter should switch antenna or not.

Alternatively, the transmitter may obtain the forward channel information based on the received signal on the reverse channel. This second approach is particularly applicable to wireless systems employing time division duplexing (TDD) scheme. Note that with TDD, the forward and reverse channels are using the same frequency channel at different time. The fading realization on forward and reverse channels may be highly correlated, if not identical, which is usually referred to as the *channel reciprocity property*. As such, the transmitter may estimate the reverse channel state information based on the pilots transmitted in the reverse channel and use the estimates as the forward channel state. With this alternative approach, the receive will transmit some pilot symbols instead of its estimated channel state information. Figure 5.15 illustrates both approaches for CSI acquisition at the transmitter.

5.5.2 Channel knowledge unavailable at transmitter

When no channel state information can be made available at the transmitter side, we can still explore the diversity benefit inherent in the multiple transmit antennas through a class of linear transmission schemes, termed as space-time coding (STC). Conventional diversity transmission, e.g., MRT, concentrates on the transmitted signal design in the spatial domain only, i.e., what signal should be transmitted from different transmit antennas over the current symbol period. STC transmission designs the transmitted signal in both spatial and temporal domains. In this subsection, we explain the basic idea of STC and complex orthogonal design.

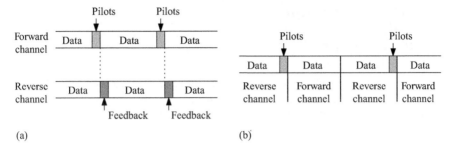

(a)   (b)

Figure 5.15   Acquiring channel state information at the transmitter: (a) FDD and (b) TDD

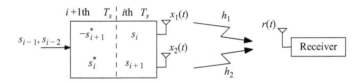

Figure 5.16   Alamouti space-time-coded transmission

We first present by far the most famous STC scheme, the Alamouti scheme. Consider the simple transmit diversity scenario with two transmit antennas and one receive antenna. The Alamouti scheme transmits two complex data symbols, denoted by s_1, s_2 for convenience, over two symbol periods using two antennas. Specifically, in first symbol period, s_1 is transmitted from antenna #1 and s_2 from antenna #2. In second symbol period, $-s_2^*$ is transmitted from antenna #1 and s_1^* from antenna #2. Essentially, each data symbol is transmitted twice over two symbol periods, as illustrated in Figure 5.16.

Assuming that channel gains from the i transmit antenna to the receive antenna, denoted by $h_i, i = 1, 2$, remain constant over two symbol periods, the received symbols are given by

$$r_1 = h_1 s_1 + h_2 s_2 + n_1;$$
$$r_2 = h_1(-s_2^*) + h_2 s_1^* + n_2, \qquad (5.49)$$

where n_1 and n_2 are additive Gaussian noise collected by the receiver. Note that the detection of s_1 and s_2 based only on r_1 or only on r_2 will have poor performance because of mutual interference. The Alamouti scheme suggests detecting s_1 or s_2 based on both r_1 and r_2 as follows. We first rewrite the received symbols, after taking the complex conjugate of r_2, into matrix form as

$$\begin{bmatrix} r_1 \\ r_2^* \end{bmatrix} = \mathbf{H} \cdot \begin{bmatrix} s_1 \\ s_2 \end{bmatrix} + \begin{bmatrix} n_1 \\ n_2^* \end{bmatrix}. \qquad (5.50)$$

where the matrix $\mathbf{H}$ is given by

$$\mathbf{H} = \begin{bmatrix} h_1 & h_2 \\ h_2^* & -h_1^* \end{bmatrix}. \tag{5.51}$$

We assume that the receiver knows the complex channel gains, h_1 and h_2, and equivalently the matrix $\mathbf{H}$, through channel estimation. The receiver will multiply the received symbols with the normalized version of the Hermitian transpose of matrix $\mathbf{H}$, given by

$$\overline{\mathbf{H}^H} = \frac{1}{\sqrt{|h_1|^2 + |h_2|^2}} \begin{bmatrix} h_1 & h_2 \\ h_2^* & -h_1^* \end{bmatrix}^H = \frac{1}{\sqrt{|h_1|^2 + |h_2|^2}} \begin{bmatrix} h_1^* & h_2 \\ h_2^* & -h_1 \end{bmatrix}, \tag{5.52}$$

where $[\cdot]^H$ denotes Hermitian transpose and performs detection on the resulting vector, given by

$$\begin{bmatrix} z_1 \\ z_2 \end{bmatrix} = \frac{1}{\sqrt{|h_1|^2 + |h_2|^2}} \begin{bmatrix} h_1^* & h_2 \\ h_2^* & -h_1 \end{bmatrix} \begin{bmatrix} r_1 \\ r_2^* \end{bmatrix}. \tag{5.53}$$

After proper substitution and carrying out matrix multiplication, we can show that

$$\begin{bmatrix} z_1 \\ z_2 \end{bmatrix} = \sqrt{|h_1|^2 + |h_2|^2} \begin{bmatrix} s_1 \\ s_2 \end{bmatrix} + \frac{1}{\sqrt{|h_1|^2 + |h_2|^2}} \begin{bmatrix} h_1^* & h_2 \\ h_2^* & -h_1 \end{bmatrix} \begin{bmatrix} n_1 \\ n_2^* \end{bmatrix}. \tag{5.54}$$

It follows that z_i can be used to detect s_i without mutual interference, since z_i is related to s_i as

$$z_i = \sqrt{|h_1|^2 + |h_2|^2} s_i + \tilde{n}_i, \quad i = 1, 2, \tag{5.55}$$

where $\tilde{n}_i$ denote the noise term, given by

$$\tilde{n}_1 = \frac{h_1^* n_1 + h_2 n_2^*}{\sqrt{|h_1|^2 + |h_2|^2}}, \quad \tilde{n}_2 = \frac{h_2^* n_1 - h_1 n_2^*}{\sqrt{|h_1|^2 + |h_2|^2}}. \tag{5.56}$$

It can be shown that noise term $\tilde{n}_i$ is zero-mean Gaussian with variance N_0. As such, the effective SNR for the detection of s_i based on z_i is

$$\gamma = (|h_1|^2 + |h_2|^2) \frac{E_s}{2N_0} = \frac{1}{2}(\gamma_1 + \gamma_2). \tag{5.57}$$

Therefore, the Alamouti scheme achieves similar diversity benefit as transmit MRC, while only requiring that the receiver knows the channel gain information. The transmitter does not need any knowledge about the channel. While different information symbols are simultaneously transmitted from two antennas, they will not interfere with each other after the linear processing on received symbols over two symbol periods. There is a 3-dB power loss compared to transmit MRC scheme due to the fact that each symbol is transmitted twice over two symbol periods. Another attractive feature of the Alamouti scheme is that two data symbols are transmitted over two symbol periods. Therefore, the system suffers no rate loss compared to conventional transmission systems.

The generalization of space-time-coded transmission to more than two transmit antenna cases has also been investigated. The Alamouti scheme is able to combine coding at the transmitter and linear processing at the receiver because of the existence of the matrix

$$\mathbf{H} = \begin{bmatrix} h_1 & h_2 \\ h_2^* & -h_1^* \end{bmatrix}, \tag{5.58}$$

which satisfies that

$$\mathbf{H}^H \mathbf{H} = (|h_1|^2 + |h_2|^2)\mathbf{I}, \tag{5.59}$$

where $\mathbf{I}$ is the 2-by-2 identity matrix. As such, considerable efforts have been carried out to construct complex matrix with such property for multiple transmit antenna scenario. These efforts lead to the theory of generalized complex orthogonal design. Specifically, for space time coded transmission over L transmit antennas, we needs to design an M-by-L matrix $\mathbf{G}$ with entries belonging to the set $\{0, \pm s_1, \pm s_1^*, \pm s_2, \pm s_2^*, \ldots, \pm s_N, \pm s_N^*\}$ and satisfying the relationship

$$\mathbf{G}^H \mathbf{G} = \left(\sum_{j=1}^{l} |s_N|^2 \right) \mathbf{I}. \tag{5.60}$$

Here, matrix $\mathbf{I}$ is the L-by-L identity matrix. It has been shown that such $\mathbf{G}$ matrix exists for arbitrary L, but with the number of row $M \geq L$ and $M \geq N$. A sample design for $L = 4$ is given by

$$\mathbf{G}_4 = \begin{bmatrix} s_1 & s_2 & s_3 & s_4 \\ -s_2 & s_1 & -s_4 & s_3 \\ -s_3 & s_4 & s_1 & -s_2 \\ -s_4 & -s_3 & s_2 & s_1 \\ s_1^* & s_2^* & s_3^* & s_4^* \\ -s_2^* & s_1^* & -s_4^* & s_3^* \\ -s_3^* & s_4^* & s_1^* & -s_2^* \\ -s_4^* & -s_3^* & s_2^* & s_1^* \end{bmatrix} \tag{5.61}$$

Note that with $\mathbf{G}_4$, $N = 4$ data symbols are transmitted over $M = 8$ symbol periods using $L = 4$ antennas. While achieving full diversity benefit, the system suffers a rate loss of a half compared with conventional transmission systems. The Alamouti design for $L = 2$ is the only space time coding scheme with rate N/M equal to one. As a result, the Alamouti scheme has been included in several wireless standards.

5.6   Further readings

The performance analysis of these conventional diversity combining schemes over more general fading environment can be found in [1]. Reference [2, Chapters 4 and 5] presents several advanced diversity combining techniques. The generalization of the diversity combining scheme over multi-transmit-multi-receive-antenna scenario will be presented in later chapter. Further discussion about space-time-coded wireless transmissions can be found in [3].

Problems

1. The maximal tolerable instantaneous BER of a digital wireless transmission system with QPSK is 10^{-4}. L-branch SC is implemented at the receiver and the system is operating over i.i.d. Rayleigh fading channels. Determine the average received SNR that the system needs to maintain at the receive antenna to satisfy the outage probability requirement of 0.05 when (i) $L = 1$; (ii) $L = 2$; and (iii) $L = 3$.
2. Derive the closed-form expression of the average BER of BPSK with L-branch SC over i.i.d. Rayleigh fading channels.
3. Repeat Problem 1 assuming L-branch MRC scheme is adopted.
4. Apply the MGF-based approach to derive an analytical expression for the average error rate of 16-QAM with L-branch MRC receiver over i.i.d. Nakagami fading channels.
5. Sketch the structure of an EGC-based diversity receiver.
6. Derive the average error rate of binary DPSK with L-branch SEC receiver over i.i.d. Rayleigh fading channels.
7. The channel knowledge at the transmitter side is usually obtained through feedback from the receiver. Assume that the channel coherence time of a wireless channel is 0.2 ms. Determine the feedback load of the following systems in bits/second: (i) selection-based transmit diversity system with $L = 4$; (ii) transmit diversity system based on Alamouti scheme with $L = 2$; and (iii) MRC-based transmit diversity with $L = 2$ antennas. Assume the real and imaginary parts of the channel gains are each quantized to 8 bits.
8. Consider the following two space-time block code designs for four transmit antenna case.

$$C_4 = \begin{bmatrix} c_1 & c_2 & c_3 & 0 \\ -c_2^* & c_1^* & 0 & c_3 \\ -c_3^* & 0 & c_1^* & -c_2 \\ 0 & -c_3^* & c_2^* & c_1 \end{bmatrix} \tag{5.62}$$

and

$$C_{4,1} = \begin{bmatrix} c_1 & c_2 & c_3 & c_4 \\ -c_2^* & c_1^* & -c_4^* & c_3^* \\ -c_3^* & -c_4^* & c_1^* & c_2^* \\ c_4 & -c_3 & -c_2 & c_1 \end{bmatrix} \tag{5.63}$$

Determine the rate of each code and which one achieves full diversity gain without intersymbol interference.

Bibliography

[1]   M. K. Simon and M.-S. Alouini, *Digital Communication over Fading Channels*, 2nd ed. New York, NY: John Wiley & Sons, 2005.

[2]   H.-C. Yang and M.-S. Alouini, *Order Statistics in Wireless Communications*, New York, NY: Cambridge University Press, 2011.

[3]   A. Paulraj, R. Nabar, and D. Gore, *Introduction to Space-Time Wireless Communications*, New York, NY: Cambridge University Press, 2003.

Chapter 6
Transmission over frequency-selective fading

The wireless channel introduces frequency-selective fading to the transmitted signal when the channel coherence bandwidth B_c is smaller than transmitted signal bandwidth B_s. In this chapter, we discuss the challenges and candidate solutions for digital wireless transmission over frequency-selective fading channels. We first discuss the effects of selective fading on digital wireless transmission. We then present two classes of transmission technologies for frequency-selective fading channels, namely equalization and multicarrier transmission. Special emphasis was placed on orthogonal frequency division multiplexing (OFDM) technology, the practical discrete implementations of multicarrier transmission, as several advanced wireless systems adopt this transmission technology.

6.1 Effect of frequency-selective fading

Previous chapters concentrate flat fading channels, where the effect of the channel is characterized by a complex channel gain z. Such scenario occurs when the channel delay spread is negligible compared to symbol period, i.e., $\sigma_T \ll T_s$. To satisfy the growing demand for high data rate wireless services, wireless communication systems use increasingly smaller symbol period or equivalently, larger channel bandwidth. For example, the typical channel bandwidth of first-generation cellular systems is 30 kHz, whereas the signal bandwidth of most popular second-generation cellular system, Global System for Mobile communications (GSM), is 200 kHz. The signal bandwidth increased to 1–2 MHz in the third-generation cellular systems and as high as 20 MHz in the fourth-generation systems. Meanwhile, the channel delay spread σ_T and the corresponding channel coherence bandwidth T_c depend on radio propagation environment and remain more or less unchanged. As such, the channel delay spread becomes significant compared to symbol period, i.e., $\sigma_T \gtrsim T_s$, or equivalently, the channel coherence bandwidth B_c is smaller than transmitted signal bandwidth B_s. The transmitted signals of these wireless systems will experience frequency-selective fading.

The received complex baseband signal over frequency-selective fading channels is given by

$$v(t) = \sum_{n=1}^{N} \alpha_n e^{j\phi_n} u(t - \tau_n) + \tilde{n}(t), \tag{6.1}$$

where $\alpha_n e^{j\phi_n}$ is the complex gain and τ_n is the delay, both for the nth path, $u(t)$ is the transmitted complex baseband signal, and $\tilde{n}(t)$ is the complex baseband additive noise. Essentially, the channel acts as a discrete filter with impulse response given by

$$h(t) = \sum_{n=1}^{N} \alpha_n e^{j\phi_n} \delta(t - \tau_n), \tag{6.2}$$

where $\delta(\cdot)$ is the unit impulse signal. As such, the channel will modify the spectrum of transmitted signals. In particular, the spectrum of the received complex baseband signal $v(t)$ is determined as

$$V(f) = \mathscr{F}\{v(t)\} = \mathscr{F}\{u(t) * h(t) + \tilde{n}(t)\} = H(f) \times U(f) + N(f), \tag{6.3}$$

where $\mathscr{F}\{\cdot\}$ denotes the Fourier transform operation, $U(f)$ is the spectrum of $u(t)$, $H(f)$ is the channel frequency response, and $N(f)$ is the noise spectrum. The spectrum of received radio frequency (RF) signal $r(t)$ is obtained by shifting that of $V(f)$ to the carrier frequency f_c, noting the relationship $r(t) = \mathrm{Re}\{v(t)e^{j2\pi f_c t}\}$.

The filtering effect of the wireless channel is illustrated in Figure 6.1. Here, the spectrum of transmitted signal is assumed to be constant over the signal bandwidth B_s. The wireless channel introduces frequency-selective fading when the channel coherence bandwidth B_c is less than B_s. As such, the frequency response of the channel $H(f)$ varies considerably over the signal bandwidth B_s, as shown in Figure 6.1. The spectrum of received signal is equal to the spectrum of the transmitted signal multiplied with the channel frequency response. Note that the spectrum of the received signal becomes much different from that of the transmitted signal. As such, the direct detection of transmitted signal based on the received signal will lead to poor performance, even when the noise component is negligible.

The received RF signal over selective fading can be written as

$$r(t) = \mathrm{Re}\left\{ \sum_{n=1}^{N} \alpha_n e^{j\phi_n} u(t - \tau_n) e^{j2\pi f_c(t-\tau_n)} \right\} + n(t), \tag{6.4}$$

where $n(t)$ is the additive white Gaussian noise (AWGN).

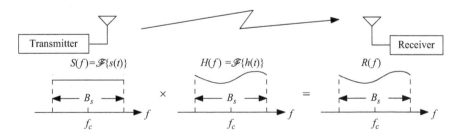

Figure 6.1   The filtering effect of selective fading channel

Applying the discrete-time approximate model for selective fading channels developed in Chapter 3, $r(t)$ can be rewritten as

$$r(t) = \text{Re}\left\{\sum_{l=1}^{L} h_l u(t - l\Delta\tau)e^{j2\pi f_c(t - l\Delta\tau)}\right\} + n(t), \tag{6.5}$$

where h_l denotes the gain of the lth tap. For digital wireless transmission, a data symbol s_i was transmitted over each symbol period after proper pulse shaping. The transmitted complex baseband signal is given by

$$u(t) = \sum_{i=-\infty}^{+\infty} s_i g(t - iT_s), \tag{6.6}$$

where s_i is the transmitted symbol over ith symbol period and $g(t)$ is the pulse shape. If we set $\Delta\tau = T_s$, the received complex baseband signal is given by

$$v(t) = \sum_{l=0}^{L-1} h_l \left(\sum_{i=-\infty}^{+\infty} s_i g(t - lT_s - iT_s)\right) + \tilde{n}(t). \tag{6.7}$$

We assume that the receive applies matched filter detection with filter response $g^*(T_s - t)$. Furthermore, $g(t)$ is properly chosen such that the overall response of pulse shaping and matched filter, $g(t) \times g^*(T_s - t)$, satisfies the Nyquist criterion. It can be shown, while referring to the illustration in Figure 6.2, that the received baseband symbol over the ith symbol period is given by

$$r_i = \sum_{l=0}^{L-1} h_l s_{i-l} + n_i. \tag{6.8}$$

In particular, the received symbol becomes the linear convolution of the transmitted symbols with the channel tap gains plus additive noise sample n_i. The effect of the composite baseband channel, consisting of transmit pulse shaping/RF front end, selective fading channel, and receiver RF front end/matched filtering, can be characterized by the discrete channel response h_l, $l = 0, 1, 2, \ldots, L - 1$. Here, L is often referred to as the length of the channel.

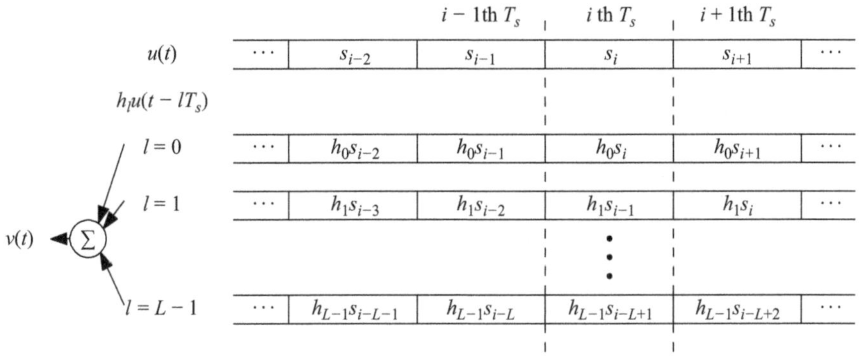

		$i - 1$th T_s	ith T_s	$i + 1$th T_s	
$u(t)$	$\cdots$	s_{i-2}	s_{i-1}	s_i	s_{i+1}	$\cdots$
$h_l u(t - lT_s)$						
$l = 0$	$\cdots$	$h_0 s_{i-2}$	$h_0 s_{i-1}$	$h_0 s_i$	$h_0 s_{i+1}$	$\cdots$
$l = 1$	$\cdots$	$h_1 s_{i-3}$	$h_1 s_{i-2}$	$h_1 s_{i-1}$	$h_1 s_i$	$\cdots$
$l = L - 1$	$\cdots$	$h_{L-1} s_{i-L-1}$	$h_{L-1} s_{i-L}$	$h_{L-1} s_{i-L+1}$	$h_{L-1} s_{i-L+2}$	$\cdots$

$v(t) \to \Sigma$

Figure 6.2 Intersymbol interference due to selective fading

The received symbol can be rewritten as

$$r_i = h_0 s_i + \sum_{l=1}^{L-1} h_l s_{i-l} + n_i.$$ (6.9)

If one tried to detect the transmitted symbol s_i, then the term $\sum_{l=1}^{L-1} h_l s_{i-l}$ will act as interference. Such interference from previously transmitted symbols, usually referred to as *intersymbol interference (ISI)*, is inherent to frequency-selective fading channels. If not properly mitigated, ISI will severely degrade the performance of digital wireless transmission. The signal-to-interference-plus-noise ratio (SINR) for the detection of s_i based on r_i is given by

$$\gamma = \frac{|h_0|^2 E_s}{\sum_{i=1}^{L-1} |h_i|^2 E_s + N_0},$$ (6.10)

where E_s is symbol energy and N_0 is the noise power spectrum density. Note that increasing the transmission power can slightly improve the SINR but cannot effectively reduce the effect of ISI.

Example: Bit error probability over selective fading channels

Consider a digital transmission system over a wireless channel using BPSK modulation scheme. The transmission power is 30 mW, the bandwidth of the transmitted signal is 30 kHz, and the adopted pulse-shaping function leads to $B_s T_s = 1$. The additive white Gaussian noise at the receiver has a power spectrum density of 10^{-4} mW/Hz. Assume that the wireless channel introduces frequency-selective fading with instantaneous tapped-delay channel model specified as $L = 3$, $h_0 = 0.3 e^{j3\pi/5}$, $h_1 = 0.15 e^{j7\pi/4}$, and $h_2 = 0.08 e^{j\pi/3}$. Determine the instantaneous error rate of the transmission. What if the transmission power is increased to 60 mW?

Solutions: Noting that the transmit power $P_s = E_s/T_s = B_s E_s = 30$ mW, it follows that the transmitted symbol energy is calculated as

$$E_s = \frac{P_s}{B_s} = \frac{30 \times 10^{-3}}{30 \times 10^3} = 10^{-6} \text{ J.}$$ (6.11)

The received SINR can then be determined as

$$\gamma = \frac{0.3^2 \times 10^{-6}}{0.15^2 \times 10^{-6} + 0.08^2 \times 10^{-6} + 10^{-7}} = 2.31.$$ (6.12)

The corresponding instantaneous bit error probability for BPSK modulation is $Q(\sqrt{2 \cdot 2.31}) = 1.78 \times 10^{-2}$. When $P_s = 60$ mW, we have $E_s = 2 \times 10^{-6}$, which leads to a SINR of 2.65. The corresponding BER becomes $Q(\sqrt{2 \cdot 2.65}) = 1.07 \times 10^{-2}$. The performance improvement is limited because the power of ISI also increases with the transmission power.

6.2   Equalization

Equalization generally refers to various signal processing techniques applied at the receiver to mitigate the effect of frequency-selective fading channel. Due to the channel filtering effect, direct detection based on received signal will lead to poor performance. The basic idea of equalization is to preprocess the received signal, with the knowledge of the channel response, to remove the channel filtering effect and then perform detection. In this section, we present the principle of equalization and then discuss its sample digital implementations.

6.2.1   Equalizing receiver

Equalizing receiver preprocesses the received signal in order to remove the effect of the selective fading channel. Figure 6.3 presents a receiver structure with equalization. In particular, the receiver implements a preprocessing filter, usually referred to as *equalizer*. The frequency response of basic analog equalizer is designed to be the inverse of the channel frequency response over the signal bandwidth, i.e.,

$$H_{eq}(f) = \frac{c}{H(f)}, \tag{6.13}$$

where c is a normalizing constant, depending on the output power constraint of the equalizer. As a result, the overall frequency response of the channel and the preprocessing filter becomes flat over the signal bandwidth, hence the name "equalization." The spectrum of the signal component at the equalizer output will have the same shape as that of the transmitted signal, and as such, the channel filtering effect is successfully removed. It is important to note that the equalizer design requires the

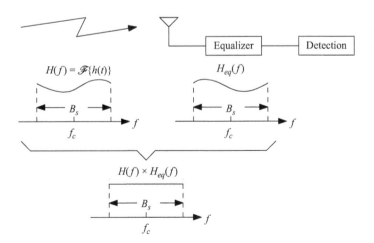

Figure 6.3   The filtering effect of selective fading channel

accurate knowledge of the channel response, which can be obtained at the receiver through channel estimation.

While capable of eliminating the effect of selective fading on received signal, the equalizer will necessarily introduce a serious side effect, known as *noise coloring/enhancement*. The receiver front end will collect additive white Gaussian noise in addition to the received signal, both of which will be processed by the equalizer. The spectrum of the complex baseband signal at the equalizer output is given by

$$Y(f) = H_{eq}(f) \times V(f) = c \times U(f) + H_{eq}(f) \times N(f). \tag{6.14}$$

The spectrum of the additive noise is modified by the filter, resulting a colored Gaussian noise with power spectral density proportional to $|H_{eq}(f)|^2$. Specifically, if the noise spectral density of white noise $N(f)$ is $N_0/2$, then that of the colored noise $N'(f) = H_{eq}(f) \times N(f)$ is given by $c^2 N_0/(2|H(f)|^2)$. If $|H(f)|$ is small at certain frequencies, then the equalizing filter will greatly enhance the noise power at those frequencies. Such noise coloring/enhancement may seriously affect the detection performance of the receiver.

The noise coloring/enhancement effect may be avoided by applying equalization at the transmitter side, which is typically referred to as transmitter pre-equalization. Essentially, the transmitter will multiply $H_{eq}(f)$ to the transmitted signal before transmission. The spectrum of the complex baseband transmitted signal will be $H_{eq}(f) \times U(f)$ and that of the received signal is calculated as

$$V(f) = H(f) \times H_{eq}(f) \times U(f) + N(f) = c \times U(f) + N(f). \tag{6.15}$$

On the other hand, implementing equalizer at the transmitter will necessarily require the channel knowledge at the transmitter side, which can be challenging to realize. Furthermore, when $|H(f)|$ is small at certain frequencies, the transmitter will allocate more transmit power to those frequencies. The value of constant c will be relatively small, which leads to low received signal-to-noise ratio (SNR).

Example: Noise enhancement due to equalization

Consider a wireless channel with frequency response over the signal bandwidth given by

$$H(f) = 1/\sqrt{|f - f_c| + 5}, \quad |f - f_c| \le B_s/2. \tag{6.16}$$

Assume that the channel bandwidth B_s is 30 kHz and the noise spectrum density at the receiver front end is $N_0 = 10^{-4}$ mW/Hz. Determine the noise power before and after the equalizer with $c = 1$.

Solutions: The noise power before equalization is simply $B_s N_0 = 3$ mW. The power spectrum of the noise signal after equalization is $N_0/(2|H(f)|^2) = 0.5N_0(|f - f_c| + 5)$. Therefore, the noise power can be calculated as

$$N = \int_{f_c - B_s/2}^{f_c + B_s/2} 0.5 N_0(|f - f_c| + 5) df = (B_s^2/4/2 + B_s/2)N_0 = 113 \text{ mW}. \quad (6.17)$$

The noise power is increased dramatically by the equalization operation.

6.2.2 Digital implementation

Most modern wireless transmission systems typically implement the equalizing operation digitally. The equalizer designs are based on the complex baseband input/output relationship for selective fading channels that we developed in previous section. In particular, the received symbol over the ith symbol period is given by

$$r_i = \sum_{l=0}^{L-1} h_l s_{i-l} + n_i = h_0 s_i + \sum_{l=1}^{L-1} h_l s_{i-l} + n_i, \quad (6.18)$$

where s_i is the transmitted symbols and $h_l, l = 0, 1, 2, \ldots, L - 1$, characterize the channel response. Note that the received symbol over the current symbol period involves the transmitted over the current and previous $L - 1$ symbol periods. If not properly mitigated, the ISI term $\sum_{l=1}^{L-1} h_l s_{i-l}$ will negatively affect the detection performance and result an irreducible floor of the average bit error rate, even when the noise power becomes negligible. Many equalizer structures have been proposed for digital receiver in presence of ISI. In what following, we present several fundamental linear equalizers.

Linear equalizers are typically implemented as an N-tap transversal filter, as shown in Figure 6.4. Specifically, the equalizer output over the ith symbol period is generated using current and previous $N - 1$ received symbols as

$$y_i = \sum_{n=0}^{N-1} w_n r_{i-n}, \quad (6.19)$$

where $w_n, n = 0, 1, 2, \ldots, N - 1$, are the tap weights. Different equalizers determine and update the tap weights for different design goals. With *zero-forcing (ZF)* equalizer, the tap weights are chosen to force ISI equal to zero. In particular, w_ns are chosen such as

$$y_i \approx s_i + \hat{n}_i, \quad (6.20)$$

where $\hat{n}_i$ is the output noise. Comparing with the result in (6.19) after substituting (6.18), we can show that the equalizer weights should satisfy

$$[h_0, h_1, \ldots, h_{L-1}] * [w_0, w_1, \ldots, w_{N-1}] = [1, 0, 0, \ldots], \quad (6.21)$$

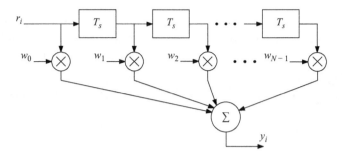

Figure 6.4   The structure of a linear equalizer

where $*$ denotes the convolution operation on two sequences. Applying Z-transform, we can show that the desired system function of ZF equalizer should be

$$H_{zf}(z) = \frac{1}{H(z)} = \frac{1}{h_0 + h_1 z^{-1} + \cdots + h_{L-1} z^{-(L-1)}}. \tag{6.22}$$

Unfortunately, such system function cannot be implemented exactly with finite-length linear transversal filter. Note that the system function of N-tap transversal filter is of the form

$$H_{trans}(z) = w_0 + w_1 z^{-1} + \cdots + w_{N-1} z^{-(N-1)}. \tag{6.23}$$

As such, equalizer with finite-length transversal filter implementation cannot completely eliminate the ISI. We can only design w_n's to approximate $1/H(z)$.

A sample method for determining the tap weights for ZF equalizer is present in the following. Assume that an N-tap transversal filter is to be designed. Given the available degree of freedom, we can design w_ns such that the first N entries of the overall impulse response of the channel and the equalizer are of the form $[1, 0, 0, \ldots, 0]$, which we denote by a $N \times 1$ column vector $\mathbf{q}$. While the remaining values of the overall response are unrestricted, it can be shown that if the first sample of the channel response h_0 is larger than the sum of the remaining samples h_l, $l = 1, 2, \ldots, L - 1$, the resulting design will minimize the maximum possible distortion due to ISI. We form an N-by-N matrix using the channel vector $\mathbf{h} = [h_0, h_1, h_2, \ldots, h_{L-1}]^T$ as

$$\mathbf{G} = \begin{bmatrix} h_0 & h_1 & \cdots & h_{N-1} \\ 0 & h_0 & \cdots & h_{N-2} \\ \vdots & \vdots & \ddots & \vdots \\ 0 & 0 & \cdots & h_0 \end{bmatrix}. \tag{6.24}$$

The tap weights in a vector form, denoted by $\mathbf{w} = [w_0, w_1, \ldots, w_{N-1}]^T$, are then determined as

$$\mathbf{w} = (\mathbf{G}^{-1})^T \mathbf{q}. \tag{6.25}$$

In other word, $\mathbf{w}$ for ZF equalizer is the first row of the inverse matrix of $\mathbf{G}$.

Digital ZF equalizer suffers the same noise coloring/enhancement problem of the analog equalizer in previous section. The *minimum mean-square error (MMSE)* equalizer achieves better performance than the ZF equalizer by properly balancing ISI mitigation and noise enhancement. The goal of MMSE equalizer is to minimize the mean square error (MSE) between the transmitted symbol and equalizer output, which is mathematically defined as

$$J = \mathbf{E}[|y_i - s_i|^2], \tag{6.26}$$

where $\mathbf{E}[\cdot]$ represents the statistical expectation operation. Note that y_i can be written in the form of the inner product of two vectors as

$$y_i = \mathbf{w}^T \mathbf{r}, \tag{6.27}$$

where $\mathbf{r} = [r_i, r_{i-1}, \ldots, r_{i-N+1}]^T$ is the vector of received symbols. Therefore, the MSE can be rewritten as

$$\begin{aligned} J &= \mathbf{E}[\mathbf{w}^T \mathbf{r} \mathbf{r}^H \mathbf{w}^* - 2\text{Re}\{\mathbf{r}^H \mathbf{w}^* s_i\} + |s_i|^2] \\ &= \mathbf{w}^T \Phi_r \mathbf{w}^* - 2\text{Re}\{\varphi_{rs} \mathbf{w}^*\} + \mathbf{E}[|s_i|^2], \end{aligned} \tag{6.28}$$

where $\Phi_r = \mathbf{E}[\mathbf{r}\mathbf{r}^H]$ denotes the autocorrelation matrix of $\mathbf{r}$ and $\varphi_{rs} = \mathbf{E}[\mathbf{r}^H s_i]$ is the cross-correlation vector between $\mathbf{r}$ and s_i.

Since J is a quadratic function of weight vector $\mathbf{w}$, we can determine the optimal weights that minimize the MSE by setting the gradient equal to 0. Specifically, taking the derivatives of J with respect to $\mathbf{w}$ leads to

$$\frac{dJ}{d\mathbf{w}} = 2\mathbf{w}^T \Phi_r - 2\varphi_{rs}. \tag{6.29}$$

Setting $\frac{dJ}{d\mathbf{w}}$ to zero, we can solve the MMSE equalizer weights as

$$\mathbf{w}_{\text{opt}} = (\Phi_r^T)^{-1} \varphi_{rs}^T. \tag{6.30}$$

The autocorrelation matrix Φ_r is in general given by

$$\Phi_r = \begin{bmatrix} \mathbf{E}[r_i r_i^*] & \mathbf{E}[r_i r_{i-1}^*] & \cdots & \mathbf{E}[r_i r_{i-N+1}^*] \\ \mathbf{E}[r_{i-1} r_i^*] & \mathbf{E}[r_{i-1} r_{i-1}^*] & \cdots & \mathbf{E}[r_{i-1} r_{i-N+1}^*] \\ \vdots & \vdots & \ddots & \vdots \\ \mathbf{E}[r_{i-N+1} r_i^*] & \mathbf{E}[r_{i-N+1} r_{i-1}^*] & \cdots & \mathbf{E}[r_{i-N+1} r_{i-N+1}^*] \end{bmatrix}. \tag{6.31}$$

For independent unit-energy data symbols and AWGN with normalized spectrum density $N_0' = N_0/E_s$, we can show

$$\mathbf{E}[r_{i-\mu} r_{i-\lambda}^*] = \begin{cases} \sum_{l=0}^{L-1} |h_l|^2 + N_0', & \lambda = \mu; \\ \sum_{l=0}^{L-1-(\lambda-\mu)} h_l h_{l+\lambda-\mu}^*, & \lambda > \mu; \\ \sum_{l=0}^{L-1-(\mu-\lambda)} h_{l+\mu-\lambda} h_l^*, & \lambda < \mu. \end{cases} \tag{6.32}$$

Similarly, the cross-correlation vector φ_{rs} is given, assuming zero equalizion delay, by

$$\varphi_{rs} = \begin{bmatrix} h_0 & 0 & \cdots & 0 \end{bmatrix}. \tag{6.33}$$

We can see that the MMSE equalizer weights are determined while taking into account the AWGN noise. MMSE equalizer in general outperforms ZF equalizer but has higher computational complexity. Note that solving for optimal weights $\mathbf{w}_{opt}$ requires the inversion of the autocorrelation matrix Φ_r.

Example: Comparison of ZF and MMSE equalizers

A selective fading channel has the following channel vector $\mathbf{h} = [0.8, -0.2, 0.14, 0.1, -0.05]$. The normalized noise spectrum density N_0/E_s is 0.02. Determine the tap weights for

1. a 3-tap ZF equalizer and
2. a 3-tap MMSE equalizer for this channel.

Solutions:

1. Given the channel vector, the target overall response should be $[1, 0, 0]$ for 3-tap ZF equalizer. The $\mathbf{G}$ matrix is constructed as

$$\mathbf{G} = \begin{bmatrix} 0.8 & -0.2 & 0.14 \\ 0 & 0.8 & -0.2 \\ 0 & 0 & 0.8 \end{bmatrix}. \tag{6.34}$$

   The tap weights for ZF equalizer are determined as $[1.25, 0.31, -0.14]$.

2. The autocorrelation matrix and cross-correlation vector for received signal over the channel are determined as

$$\Phi_r = \begin{bmatrix} 0.73 & -0.18 & 0.09 \\ -0.18 & 0.73 & -0.18 \\ 0.09 & -0.18 & 0.73 \end{bmatrix} \tag{6.35}$$

   and

$$\varphi_{rs} = \begin{bmatrix} 0.8 & 0 & 0 \end{bmatrix}. \tag{6.36}$$

   The equalizer weights for MMSE are determined as $[1.17, 0.27, -0.07]$. Note that when the noise power approaches zero, the tap weights for MMSE equalizer become $[1.20, 0.28, -0.07]$, which is similar to that of ZF equalizer.

Other equalization schemes include decision-feedback equalization (DFE) and maximum likelihood sequence estimation (MLSE). DFE is a nonlinear equalizing solution that can greatly reduce the noise enhancement effect. The basic idea is to remove ISI of earlier transmitted symbols using the detection results of these symbols. A sample structure is shown in Figure 6.5. Such approach may suffer from error propagation when symbol detection is not perfect. MLSE estimates the sequence of transmitted symbols following the maximum likelihood principle. While achieving the optimal performance, such solution tends to have high computational complexity. Fortunately, the Viterbi algorithm, originally targeted at the optimal decoding of

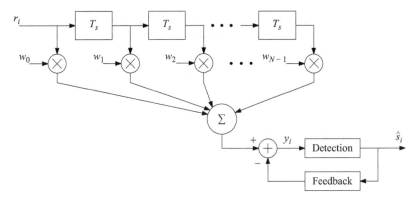

Figure 6.5   The structure of a decision feedback equalizer

convolution codes, can efficiently achieve MLSE over frequency-selective fading channels.

6.2.3   Adaptive implementation

The equalizer design that we presented in previous subsection assumes that the channel response is known to the receiver and remains unchanged for the duration of transmission. In practice, the channel knowledge is not readily available and has to be obtained through channel estimation. Furthermore, the channel response will change over time for wireless transmission and some wireline transmission, e.g., power-line communication systems. Therefore, the equalizer implementation in such scenarios will involve periodic equalizer training/update, also termed as *adaptive equalization*.

Equalizer training/update should be performed at least every channel coherence time T_c. Typically, the transmitter will send some training sequence, i.e., data symbols known to the receiver, at the beginning of each training cycle. The receiver will use the corresponding received symbols to train or update the equalizer. Various signal processing algorithms can apply, leading to different tradeoffs between performance and converging speed. In particular, weight updating using least mean square algorithm results lower complexity than weight calculation using Weiner filter, but with not as good performance. It is worth noting that certain equalizer design does not require channel knowledge, as such referred to as *blind equalization*. Such approach uses decoded data symbols for training/update purpose, while certainly more efficient, blind equalizers tend to suffer error propagation.

The time duration required for training symbol transmission, if needed, and equalizer training/update should be much smaller than the channel coherence time T_c. Otherwise, the system will have poor transmission efficiency due to the excess overhead for adaptive equalization. The number of training symbols to be transmitted over each T_c depends on the number of equalizer weights to be determined/updated as well as the updating algorithms adopted. Meanwhile, different training/update algorithms require different amount of time to converge. If the channel coherence time of a selective fading channel is relatively small such that the time required for training symbol transmission is comparable with T_c, then equalization will not be the suitable countermeasure for ISI. We can apply other transmission technologies to mitigate ISI, such as multicarrier transmission.

6.3　Multicarrier transmission

Multicarrier transmission is another digital wireless transmission technology suitable for frequency-selective fading channels. The basic idea is to divide the wideband selective channel into many parallel narrowband subchannels and to transmit a low-rate substream over each subchannel. If the bandwidth of a subchannel is smaller or comparable to the channel coherence bandwidth, then each substream will experience frequency flat fading. Compared to the equalization approach, multicarrier transmission is a more proactive transmission technology for frequency-selective fading channels. Note that equalizer is trying to mitigate the negative effect of selective fading by processing the received signal, and therefore, can be deemed more of a reactive approach. By jointly designing the transmitter and receiver, multicarrier transmission ensures that the transmitted signal will not experience the negative effect of selective fading at all.

The idea of multicarrier transmission was used in certain military communication systems in the late 1950s. Conventional implementation of multicarrier transmission typically involves multiple narrowband modulator/demodulator pairs, whose high complexity greatly limits its application scenario. Multicarrier transmission starts to gain increasing popularity in emerging high data-rate wireless systems after the 1990s, mainly due to the development of its efficient discrete implementation, namely orthogonal frequency division multiplexing (OFDM). With the application of fundamental signal processing operations, OFDM achieves multicarrier transmission with a single modulator and demodulator pair. As a result, OFDM becomes the de facto choice for wireless transmission over frequency-selective fading channels.

Consider a digital wireless transmission system with symbol rate R_s. We assume that the transmitter uses raised cosine pulse shape with roll-off factor β, $0 \le \beta \le 1$, for each modulated symbol. The bandwidth of the modulated signal is then equal to

$$B_s = (1 + \beta)R_s. \tag{6.37}$$

If the coherence bandwidth of the channel B_c is less than B_s (or equivalently, the symbol period T_s is comparable or smaller than the channel RMS delay spread σ_T), then the transmitted signal will experience frequency-selective fading. Now we divide the symbol stream to be transmitted into N parallel substreams. The symbol rate of each substream becomes R_s/N and symbol period NT_s. If N is large enough such that the bandwidth of modulated substream $B_N = (1 + \beta)R_s/N$ is less than B_c (or equivalently, the symbol period NT_s is larger than σ_T), then each substream will experience frequency flat fading.

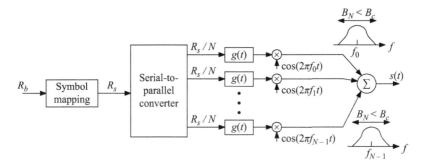

Figure 6.6   The structure of a conventional multicarrier transmitter

---

Example: Parallel decomposition of selective fading channels

A modulated data stream with symbol rate 1 Msps is being transmitted over a wireless fading channel with RMS delay spread $\sigma_T = 10\,\mu s$. The roll-off factor of pulse shaping function is 0.2. Determine the minimum number of substreams that we need to divide the stream into such that each substream will experience flat fading.

Solutions: The coherence bandwidth of the channel is estimated as $B_c = 0.2/\sigma_T = 20\,\text{kHz}$, which is much smaller than the bandwidth of transmitted signal with single-carrier transmission, equal to $(1 + 0.2) \times 1 = 1.2\,\text{MHz}$. In order to ensure that each substream experiences frequency flat fading, we need the bandwidth of each modulated substream $B_N = (1 + \beta)R_s/N$ to be less than B_c, i.e., $(1 + 0.2) \times 1 \times 10^6/N \leq 20\,\text{kHz}$. Therefore, the number of substreams should be at least 60.

Figure 6.6 illustrates the structure of a conventional multicarrier transmitter. In particular, the information bit stream is first mapped to data symbol stream, which is then converted into N parallel substreams. After proper pulse shaping, each substream is modulated onto a distinct subcarrier frequency, f_i, $i = 0, 1, \ldots, N - 1$, before being transmitted together. The carrier frequencies should be separated sufficiently apart from each other to avoid interference between substreams. Figure 6.7 shows the conventional implementation of multicarrier receiver. Bandpass filters are first applied to single out the received signal corresponding to individual substreams, before carrying out demodulation. Essentially, such transmitter/receiver structure divides wideband selective fading channel into N narrowband flat fading subchannels. The flat channel gain for each subchannel is approximately equal to the frequency response of the channel at the corresponding subcarrier frequency, i.e., $z_i = H(f_i)$, $i = 0, 1, \ldots, N - 1$.

While capable of completely eliminating ISI, multicarrier transmission usually suffers a certain bandwidth penalty, and therefore, less spectral efficient than single-carrier transmission. With the conventional multicarrier receiver structure presented

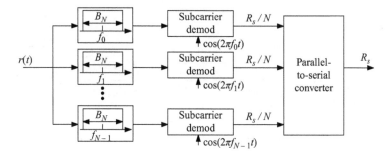

Figure 6.7   The structure of a conventional multicarrier receiver

above, additional guard bandwidths between subchannels are required to facilitate realistic bandpass filter and to accommodate practical pulse shapes. If we assume the required guard band is 100δ percent of the bandwidth of the modulated substream signal, then the total bandwidth of the transmitted signal on all substreams can be determined as

$$B_T = N(1 + \delta)B_N = (1 + \delta)B_s. \tag{6.38}$$

Here, we assume that half of the guard band is introduced to each end of the spectrum. Therefore, the multicarrier system suffers a bandwidth penalty of 100δ percent compared to conventional single-carrier transmission.

The bandwidth penalty associated with conventional multicarrier transmission originates from the requirement of nonoverlapping subchannels in the frequency domain. For digital wireless transmission with matched filter detection, the interference between substreams can be avoided as long as subcarrier frequencies are orthogonal over the subchannel symbol period $T_N = N/R_s$. In particular, the transmitted signal with conventional multicarrier transmission over a subchannel symbol period T_N is given by

$$s(t) = \sum_{j=0}^{N-1} s_j g(t) \cos(2\pi f_j t + \phi_j), \tag{6.39}$$

where s_j is the data symbol transmitted over subcarrier f_j, $g(t)$ is the pulse-shaping function, satisfying $g(t) * g^*(T_N - t) = 1$, and ϕ_j is the phase offset of the jth subcarrier. After the propagation over wireless channel, noting that the effect of the jth subchannel is the complex channel gain $z_j = H(f_j)$, the overall received signal over the symbol period of interest is given by

$$r(t) = \sum_{j=0}^{N-1} z_j s_j g(t) \cos(2\pi f_j t + \phi_j) + n(t), \tag{6.40}$$

where $n(t)$ is the additive noise signal.

Assuming coherent reception with perfect phase recovery, the sample of the matched filter output for the ith data stream is given by

$$r_i = \int_0^{T_N} r(t)g(t)\cos(2\pi f_i t + \phi_i)dt \tag{6.41}$$

$$= \int_0^{T_N} \left(\sum_{j=0}^{N-1} z_j s_j g(t)\cos(2\pi f_j t + \phi_j) \right) g(t)\cos(2\pi f_i t + \phi_i)dt + n_i,$$

where n_i is the sample of the noise component. The first term of r_i can be rewritten as

$$\sum_{j=0}^{N-1} z_j s_j \int_0^{T_N} \cos(2\pi f_j t + \phi_j)\cos(2\pi f_i t + \phi_i)dt. \tag{6.42}$$

Therefore, r_i will be equal to $z_i s_i + n_i$, i.e., suffering from no intersubstream interference, if

$$\int_0^{T_N} \cos(2\pi f_j t + \phi_j)\cos(2\pi f_i t + \phi_i)dt = 0, \ i \neq j, \tag{6.43}$$

i.e., subcarriers are orthogonal over the subchannel symbol period T_N. We can further show that the minimum frequency spacing to achieve orthogonal subcarriers is equal to $1/T_N$.

These results lead to an alternative multicarrier transmission implementation with higher spectral efficiency. The transmitter has the same structure as shown in Figure 6.6, but with subcarrier frequencies $f_i = f_0 + i/T_N$, $i = 0, 1, \ldots, N-1$. Since the bandwidth of the modulated substream signal is equal to $(1+\beta)R_s/N = (1+\beta)/T_N$, the spectrum of different substream signals will overlap in the frequency domain, as illustrated in Figure 6.8. Apparently, multicarrier transmission with overlapping subchannels achieves higher spectrum efficiency than the conventional

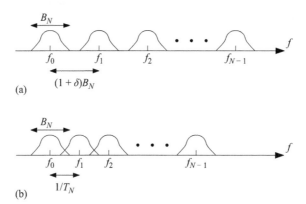

Figure 6.8   Spectrum of multicarrier modulated signals: (a) nonoverlapping subchannels and (b) overlapping subchannels

implementation with nonoverlapping subchannels. Specifically, the total required bandwidth with minimum subcarrier spacing is determined as

$$B_T = (N-1)/T_N + (1+\delta)B_N = (N-1)R_s/N + (1+\delta)(1+\beta)R_s/N, \qquad (6.44)$$

which is typically smaller than the bandwidth of single-carrier transmission $B_s = (1+\beta)R_s$.

Example: Bandwidth efficiency of multicarrier transmission

A modulated data stream with symbol rate 1 Msps is transmitted over wireless channel. To avoid intersymbol interference, multicarrier transmission with 100 subchannels is applied. The guard band between subchannels is 5% of subchannel bandwidth. The roll-off factor of pulse-shaping function is 0.2. Determine the total bandwidth requirement with

1. nonoverlapping subchannels and
2. overlapping subchannels.

Solutions:

1. The symbol rate of each subchannel is 10 ksps. Considering the guard bands, the subchannel bandwidth with nonoverlapping subchannels is $R_N(1+\beta)(1+\delta) = 10 \times 1.2 \times 1.05 = 12.6$ kHz. Therefore, the total bandwidth required is 1.25 MHz.
2. With overlapping subchannels, the subchannel spacing is 10 kHz. The total required bandwidth can be determined as $(N-1)/T_N + (1+\delta)(1+\beta)/T_N = 1.016$ MHz, which is considerably smaller than that required for nonoverlapping subchannel case. In fact, this bandwidth requirement is smaller than that of conventional single-carrier transmission, which is $R_s(1+\beta) = 1.2$ MHz.

The multicarrier receiver with matched filter detection is shown in Figure 6.9. The subcarrier frequencies are set as $f_i = f_0 + i/T_N$, $i = 0, 1, \ldots, N-1$, to avoid the intersubchannel interference. While achieving much higher spectrum efficiency than conventional non overlapping implementation, such matched filter implementation is very sensitive to timing and frequency offset between the transmitter and the receiver, which compromises the orthogonality. Nevertheless, both multicarrier implementations presented in this section requires N parallel carrier modulators and demodulators at the transmitter and the receiver, respectively, which may lead to prohibitively high system complexity and energy consumption when the number of substream required is large. In the next subsection, we present the discrete implementation of multicarrier transmission, which only requires a single modulator/demodulator pair and dramatically reduces the transceiver complexity.

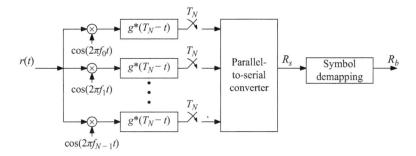

Figure 6.9 The structure of a multicarrier receiver based on matched filter detection

6.4 Discrete implementation/OFDM

The discrete implementation of multicarrier transmission eliminates the need for multiple carrier modulator/demodulator pairs with the application of fundamental digital signal processing (DSP) operations, the discrete Fourier transform (DFT), and inverse DFT (IDFT). Since both DFT and IDFT can be efficiently calculated with fast Fourier transform (FFT) algorithms, the resulting design, typically termed as OFDM, gains increasing popularity in wireless transmission systems.

Consider the transmission of N data symbols, $s_0, s_1, \ldots, s_{N-1}$, over a frequency-selective fading channel with impulse response, h_n, $n = 0, 1, \ldots, L-1$. The direct transmission of these data symbols will lead to the linear convolution of the data symbols and channel response. Each symbol will experience ISI from previously transmitted symbol. Multicarrier transmission presented in previous subsection eliminates ISI by dividing selective fading channel into parallel flat fading channels in frequency domain. Effectively, multicarrier transmission achieves the input/output relation given by

$$y_i = H(i) \cdot s_i + n_i, \quad i = 0, 1, \ldots, N-1, \tag{6.45}$$

where y_i is the decision statistics corresponding to data symbol s_i, n_i is the additive Gaussian noise, and $H(i)$ is the channel frequency response at frequency $f_i = f_0 + i/T_N$. Each transmitted data symbol will experience frequency flat fading channel with effective channel gain $H(i)$. Mathematically, $H(i)$ can be determined as the ith sample of the N-point DFT of the channel impulse response h_n, as

$$H(i) = \text{DFT}\{h_n\} = \sum_{n=0}^{L-1} h_n e^{-j2\pi ni/N}, \quad i = 0, 1, \ldots, N-1. \tag{6.46}$$

Let us focus on the signal component of y_i and neglect the additive noise. Applying N-point IDFT to y_i with the application of the convolution property of DFT, we have

$$r_n \triangleq \text{IDFT}\{y_i\} = \text{IDFT}\{H(i) \cdot s_i\} = h_n \circledast x_n, \tag{6.47}$$

where x_n denotes the N-point IDFT of s_i, defined as

$$x_n = \text{IDFT}\{s_i\} = \sum_{i=0}^{N-1} s_i e^{j2\pi ni/N}, \quad n = 0, 1, \ldots, N-1, \tag{6.48}$$

and $\circledast$ represents the circular convolution of two sequences, defined as

$$r_n = \sum_{k=0}^{L-1} h_k x_{(n-k)_N}, \tag{6.49}$$

with $(n-k)_N$ denoting $n - k$ modulo N. Essentially, r_n becomes the circular convolution of the IDFT of data symbols with the channel response h_n. From the above analysis, we arrive with the basic idea of the discrete implementation of multicarrier transmission/OFDM summarized as:

The desired input/output relation for multicarrier transmission can be achieved if we (i) transmit the IDFT of data symbols, x_n, instead of data symbols s_i themselves; (ii) manage to create the circular convolution of x_n and channel response h_n, r_n, at the receiver; and (iii) perform DFT on r_n.

The main challenge to implement discrete multicarrier transmission is how to create the circular convolution of the transmitted channel symbols x_n and channel response h_n at the receiver. Note that the channel output is typically the linear convolution of transmitted symbols and channel vector. In particular, when N symbols are transmitted over a channel with length L, the channel output is of length $N + L - 1$ due to linear convolution. Furthermore, if we transmit another block of N channel symbols immediately, earlier transmitted symbol block will affect the channel output corresponding to later transmitted symbol block, leading to the so-called interblock interference (IBI).

Introducing cyclic prefix can effectively address such challenge. In particular, adding cyclic prefix can create the circular convolution of two finite length sequences when they are linearly convolved together. To illustrate this, let us consider a length-N sequence, $[x_0, x_1, \ldots, x_{N-1}]$, and a length-$L$ sequence, $[h_0, h_1, \ldots, h_{L-1}]$, with $N > L$ without loss of generality. We use the first sequence to generate a new sequence, denoted by $[\tilde{x}_{-M}, \ldots, \tilde{x}_{-1}, \tilde{x}_0, \tilde{x}_1, \ldots, \tilde{x}_{N-1}]$, by adding M cyclic prefix symbols, where $M > L$. Specifically, the new length $N + M$ sequence $\tilde{x}_n$ is related to the sequence x_n as

$$\tilde{x}_n = \begin{cases} x_{N+n}, & n = -M, -M + 1, \ldots, -1; \\ x_n, & n = 0, 1, \ldots, N. \end{cases} \tag{6.50}$$

Then, the linear convolution of sequence $\tilde{x}_n$ and sequence h_n is calculated as

$$r_n = \sum_{k=0}^{L-1} h_k \tilde{x}_{n-k}, \quad n = -M, -M + 1, \ldots, 0, \ldots, N + L - 1. \tag{6.51}$$

Focusing the convolution results for the index n from 0 to $N-1$ while noting that $\tilde{x}_{n-k} = x_{(n-k)_N}$, we have

$$
\begin{aligned}
r_n &= \sum_{k=0}^{L-1} h_k \tilde{x}_{n-k} \\
&= \sum_{k=0}^{L-1} h_k x_{(n-k)_N} \\
&= h_n \circledast x_n, \quad n = 0, 1, \ldots, N-1.
\end{aligned}
\tag{6.52}
$$

Therefore, with the addition of cyclic prefix, linear convolution result contains circular convolution result.

Cyclic prefix can also help eliminate IBI. In particular, if the length of cyclic prefix added to the beginning of a channel symbol block is long enough, then IBI will only affect the reception of these cyclic prefix, not the results of circular convolution between data symbols and channel response. Referring to the illustration in the previous paragraph, IBI from previous block will affect r_n for $n = -M$, $-M+1, \ldots, -M+L-1$, but not those r_ns from $n = -M+L, \ldots, N-1$. Then, when $M \geq L$, the results of circular convolution r_n for $n = 0, 1, \ldots, N-1$ will not experience any IBI. Similarly, the IBI from the current symbol block to the next symbol block will only affect the received symbols corresponding to the cyclic prefix of the next block. The effect of cyclic prefix is illustrated in Figure 6.10.

Finally, we arrive at the discrete implementation of multicarrier transmission, often referred to as OFDM transceiver, as shown in Figure 6.11. At the transmitter side, the incoming bit stream is first mapped to data symbols according to the chosen modulation scheme. The resulting data symbol sequence $\ldots, s_{-1}, s_0, s_1, s_2, \ldots$ is processed by the serial-to-parallel converter, which essentially divides symbol sequences into symbol block of length N, e.g., $[s_0, s_1, \ldots, s_{N-1}]$, $[s_N, s_{N+1}, \ldots, s_{2N-1}]$, $\ldots$. Then, transmitter performs N-point inverse DFT on the data symbol blocks to generate the corresponding channel symbol blocks, i.e.,

$$
[s_0, s_1, \ldots, s_{N-1}] \xrightarrow{\text{IDFT}} [x_0, x_1, \ldots, x_{N-1}].
\tag{6.53}
$$

After parallel-to-serial conversion, M cyclic prefix symbols are added to each channel symbol block, leading to the transmitted symbol sequence $\ldots, x_{N-M}, \ldots, x_{N-1}, x_0$, $x_1, \ldots, x_{N-1}, x_{2N-M}, \ldots, x_{2N-1}, x_N, x_{N+1}, \ldots$. Note that $x_{N-M}, \ldots, x_{N-1}$ is the cyclic prefix for the first block and $x_{2N-M}, \ldots, x_{2N-1}$ for the second block. The resulting sequence is then transmitted after pulse shaping and carrier modulation.

We assume that the channel response of the selective fading channel is of length L, $L \leq M$. The corresponding received baseband symbols, after carrier removal and matched filter detection, are $\ldots, r'_{N-M}, \ldots, r'_{N-1}, r_0, r_1, \ldots, r_{N-1}, r'_{2N-M}, \ldots, r'_{2N-1}$, $\ldots$. Here, $r_0, r_1, \ldots, r_{N-1}$ are the circular convolution results of the channel symbol block with the channel response plus additive noise whereas $r'_{N-M}, \ldots, r'_{N-1}$ are the received symbols corresponding to the cyclic prefix, and as such, affected by the IBI from previous block (so are $r'_{2N-M}, \ldots, r'_{2N-1}$). The receiver will remove these prefix

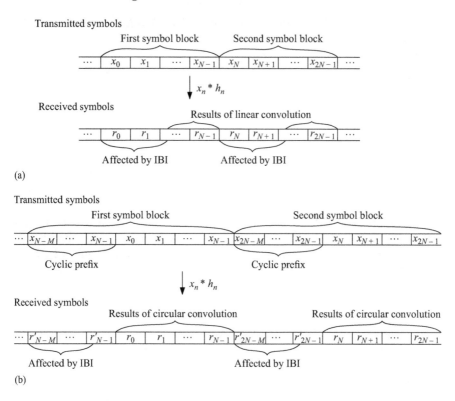

Figure 6.10   Effect of cyclic prefix on digital wireless transmission over selective fading channels: (a) without cyclic prefix and (b) with cyclic prefix

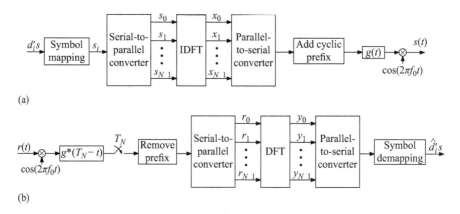

Figure 6.11   Transceiver structure for discrete implementation of multicarrier transmission: (a) transmitter and (b) receiver

$r'_{N-M}, \ldots, r'_{N-1}$ and apply N-point DFT to $r_0, r_1, \ldots, r_{N-1}$ after serial-to-parallel conversion, as

$$[r_0, r_1, \ldots, r_{N-1}] \xrightarrow{\text{DFT}} [y_0, y_1, \ldots, y_{N-1}]. \qquad (6.54)$$

Finally, we arrive at the decision statistics for the transmitted data symbol y_is, which is related to the transmitted data symbol as

$$y_i = H(i) \cdot s_i + n_i, \quad i = 0, 1, \ldots, N-1, \qquad (6.55)$$

where n_is are the DFT of the additive noise samples in $r_0, r_1, \ldots, r_{N-1}$.

The key features of the discrete multicarrier/OFDM transceiver structure include (i) IDFT operation on data symbol blocks, (ii) block channel symbol transmission with cyclic prefix insertion, and (iii) DFT operation on channel output after cyclic prefix removal. Compared to conventional single-carrier transmission, such discrete multicarrier implementation only requires an extra IDFT and DFT operation at the transmitter and receiver, respectively. With the recently development of digital technology, these operations can be efficiently performed using FFT algorithms. To facilitate the FFT calculation, the number of subcarriers or equivalently the block size is usually chosen to be a power of 2. The resulting OFDM transceiver achieves multicarrier transmission while eliminating the need for multiple carrier modulators/demodulators. As such, OFDM is becoming an increasingly popular technology for digital transmission over frequency-selective fading channels.

Example: OFDM transmission

The bit stream 1001110110001010 is being transmitted using OFDM transmitter. Quadrature phase shift keying (QPSK) modulation with bits to symbols mapping scheme given by

$$00 \Longrightarrow e^{j\frac{\pi}{4}},$$
$$01 \Longrightarrow e^{j\frac{3\pi}{4}},$$
$$11 \Longrightarrow e^{j\frac{5\pi}{4}},$$
$$10 \Longrightarrow e^{j\frac{7\pi}{4}}, \qquad (6.56)$$

is adopted. Note that symbol energy is set equal to one. Assume that the system uses four subcarriers, i.e., $N = 4$ and the channel delay spread requires two cyclic prefix symbols ($M = 2$).

1. Determine the corresponding transmitted channel symbols.
2. If the channel vector is $[0.8e^{j\frac{3\pi}{5}}, 0.3e^{j\frac{2\pi}{3}}]$, determine the signal component of decision statistics y_i.

Solutions:

1. The QPSK modulated data symbols are $e^{j\frac{7\pi}{4}}, e^{j\frac{3\pi}{4}}, e^{j\frac{5\pi}{4}}, e^{j\frac{3\pi}{4}}, e^{j\frac{7\pi}{4}}, e^{j\frac{\pi}{4}}, e^{j\frac{7\pi}{4}},$
   $e^{j\frac{7\pi}{4}}$. They are divided into two blocks of four symbols. After taking IDFT and

adding cyclic prefix, the transmitted channel symbols are $0.79e^{-j1.11}$, 0.35, $0.35e^{j3.14}$, 0.35, $0.79e^{-j1.11}$, 0.35, $0.35e^{-j1.57}$, 0.35, $0.35e^{-j0.46}$, $0.35e^{j3.14}$, $0.35e^{-j1.57}$, 0.35. Note that $0.79e^{-j1.11}$ and 0.35 are the cyclic prefix for the first block and $0.35e^{-j1.57}$, 0.35 for the second.

2. The signal component of the decision statistics y_i can be determined by first calculating the received symbols, which is the linear convolution result with the channel vector, and then perform 4-point DFT to each block after removing the prefix. The received symbols are $0.63e^{j0.78}$, $0.47e^{j1.48}$, $0.18e^{-j1.38}$, $0.18e^{j1.76}$, $0.67e^{j0.93}$, $0.47e^{j1.48}$, $0.28e^{j0.69}$, $0.32e^{j1.56}$, $0.72e^{j1.51}$, $0.08e^{-j0.41}$, $0.32e^{-j0.01}$, $0.32e^{j1.56}$, $0.11e^{j2.09}$. Note that $0.11e^{j2.09}$ will interfere the next symbol block. After removing the prefix and applying DFT on $[0.18e^{-j1.38}$, $0.18e^{j1.76}$, $0.67e^{j0.93}$, $0.47e^{j1.48}]$ and $[0.72e^{j1.51}$, $0.08e^{-j0.41}$, $0.32e^{-j0.01}$, $0.32e^{j1.56}]$, we obtain $1.10e^{j1.16}$, $0.91e^{-j2.37}$, $0.51e^{-j0.59}$, $0.79e^{-j1.66}$, $1.10e^{j1.16}$, $0.91e^{j2.34}$, $0.51e^{j0.98}$, $0.79e^{j1.48}$. Alternatively, we can obtain the same results by multiplying the 4-point DFT of the channel vector with the transmitted data symbols.

6.5   Challenges of OFDM transmission

Discrete multicarrier implementation/OFDM technology realizes multicarrier transmission with the application of IDFT and DFT operations. Meanwhile, OFDM transmission systems face several critical challenges, including the overhead of cyclic prefix, high peak-to-average power ratio, sensitivity of frequency and timing offset, and severe subcarrier fading. Some of these challenges are unique to discrete implementation while others are common to all multicarrier systems. In this subsection, we discuss these challenges and their possible mitigation schemes.

OFDM system introduces cyclic prefix to remove IBI. For the transmission of every N data symbols, the system actually transmits $N + M$ channel symbols, where M is the number of cyclic prefix symbols. The transmission of these redundant prefix symbols, although necessary, will consume extra bandwidth resource. Specifically, if the data symbol rate is R_s, then the channel symbol rate of OFDM systems will be $(N + M)R_s/N$. As such, the bandwidth requirement of OFDM system increases by a factor of $(N + M)/N$ compared to conventional transmission system.

The minimum number of cyclic prefix required is limited by the length of channel response L. Typically, given the maximum delay spread of the channel τ_{max} and the channel symbol period T_s, the length of the channel response L can be estimated as $L = \lceil \tau_{max}/T_s \rceil$. The number of cyclic prefix M should be greater than L. As such, the only way to reduce the extra spectrum usage with OFDM transmission is to increase the block size N. Meanwhile, larger N value will necessarily increase the complexity of IDFT and DFT operations and the corresponding processing delay. So, there is a tradeoff of spectral efficiency and transceiver complexity associated to OFDM transceiver design.

Example: Data rate analysis of OFDM systems

Consider the digital wireless transmission over a frequency-selective fading channel with bandwidth 1 MHz. The maximum delay spread of the channel τ_{max} is 6 μs. For the sake of simplicity, we assume that ideal pulse-shaping function with roll-off factor of $\beta = 0$ is used, i.e., $B_s \cdot T_s = 1$. Determine the largest data symbol rate of OFDM system with $N = 32$ subcarriers. What if $N = 128$ subcarriers are used instead?

Solutions: When $\beta = 0$, the channel symbol period T_s is equal to $1/B_s = 1$ μs. The duration of the cyclic prefix transmission should be longer than or equal to τ_{max}. Therefore, the smallest number of cyclic prefix symbols is $M = 6$. For OFDM system with $N = 32$, data symbols are divided into blocks of $N = 32$ symbols (32-point FFT calculation is needed at the transmitter and receiver). For each data symbol block, a total of 38 channel symbols, including the 32-point FFT of data symbols and 6 cyclic prefix, are transmitted, which takes 38 μs. The data symbol rate of the resulting system is then calculated as

$$R_s = \frac{32}{38 \times 10^{-6}} = 16/19 \text{ Msps} \tag{6.57}$$

Note that with conventional single-carrier transmission, one data symbol is transmitted over each channel symbol period T_s, which leads to the data rate of 1 Msps.

If we instead adopt a block size of $N = 128$, then the data symbol rate of the OFDM system becomes 128/134 Msps, approaching that of conventional single-carrier transmission. The transmitter and receiver will now need to perform a 128-point IDFT and DFT calculation for each data block.

Peak-to-average power ratio (PAPR), characterizing the dynamic range of the transmitted signal power, has significant effect on the power efficiency of communication systems. PAPR is defined as the ratio of the peak power over the average power of the signal. For digitally modulated signals, PAPR can be calculated as

$$\text{PAPR} \triangleq \frac{\max_n\{|x_n|^2\}}{\mathbf{E}[|x_n|^2]}, \tag{6.58}$$

where $|x_n|^2$ denotes the power of transmitted symbols and $\mathbf{E}[\cdot]$ denotes the statistical averaging operation. Here, we neglect the effect of pulse-shaping function. Based on the above definition, the PAPR of linearly modulated signal with phase shift keying modulation schemes will have a PAPR close to 1.

Meanwhile, the transmitted signal of an OFDM system typically has high PAPR, due to the inverse DFT operation at the transmitter. Note that the transmitted channel symbols of OFDM systems are determined as

$$x_n = \sum_{i=0}^{N-1} s_i e^{j2\pi ni/N}, \quad n = 0, 1, \ldots, N - 1. \tag{6.59}$$

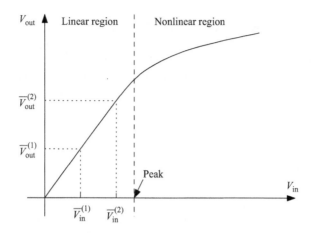

Figure 6.12   Effect of PAPR on the efficiency of transmitter power amplifier (case 1: larger PAPR; case 2: smaller PAPR)

As such, the channel symbols x_ns may have high PAPR, even when modulated data symbols s_is have identical power. Specifically, the average power of OFDM transmitted signal can be calculated, assuming independent data symbols s_i with common energy E_s, as

$$\mathbf{E}[|x_n|^2] = \sum_{i=0}^{N-1} \mathbf{E}[|s_i|^2] = NE_s. \tag{6.60}$$

The maximum value of $|x_n|^2$ occurs when $s_i e^{j2\pi ni/N}$s are added together coherently, leading to the maximum value of $\max_n\{|x_n|^2\} = (\sum_{i=0}^{N-1} |s_i|)^2$. The resulting PAPR of OFDM transmitted signal is determined as

$$\text{PAPR} = \frac{(N\sqrt{E_s})^2}{NE_s} = N. \tag{6.61}$$

While the coherent addition of all $s_i e^{j2\pi ni/N}$s rarely happens, the PAPR of OFDM signal has been shown to increase with the number of subcarriers N.

High PAPR negatively impacts the operation efficiency of transmitter power amplifier. The response curve of typical RF power amplifier is shown in Figure 6.12. To avoid signal distortion during amplification, the input signal voltage should stay in the linear region. The peak value of input signal is limited to the upper end of the linear region. As such, larger PAPR will lead to smaller average output signal voltage (see case 1). The amplifier efficiency is improved if the average value of input signal close to its peak, corresponding to small PAPR, as shown in case 2. Therefore, PAPR reduction is critical to the success of OFDM systems. Numerous research effort have been carried out to develop effective PAPR reduction solutions. Candidate solution includes signal clipping, peak cancellation with complementary signal, and dynamic null placement.

Example: PAPR of OFDM signal

Assuming the same setup as previous example on OFDM transmission, calculate the PAPR of the transmitted signal. What if the system uses $N = 8$ subcarriers instead?

Solutions: The peak power of 12 transmitted symbols, $0.79e^{-j1.11}$, 0.35, $0.35e^{j3.14}$, 0.35, $0.79e^{-j1.11}$, 0.35, $0.35e^{-j1.57}$, 0.35, $0.35e^{-j0.46}$, $0.35e^{j3.14}$, $0.35e^{-j1.57}$, 0.35, is 0.79^2. The PAPR of the transmitted symbols is calculated as

$$\text{PAPR} = \frac{0.79^2}{(2 \times 0.79^2 + 10 \times 0.35^2)/12} = 2.5. \tag{6.62}$$

Now let assume that the system uses eight subcarriers ($N = 8$). Then, the transmitted channel symbols become $0.35e^{j\pi}$, $0.33e^{j1.18}$, $0.25e^{-j0.79}$, $0.57e^{-j1.8}$, 0, $0.14e^{j2.75}$, $0.56e^{-j1.24}$, $0.23e^{j1.01}$, $0.35e^{j\pi}$, $0.33e^{j1.18}$. Note that $0.35e^{j\pi}$ and $0.33e^{j1.18}$ are the cyclic prefix, leading to a total of ten channel symbols. The PAPR of transmitted signal becomes 2.6. Apparently, larger N leads to more efficient transmission, but larger PAPR in ODFM systems.

OFDM technology achieves orthogonal multicarrier transmission through discrete implementation. When the block size N is used, the subcarrier spacing is $B_s/N = 1/T_N$. As such, neighboring subchannels are overlapping, which requires perfect orthogonality to eliminate intersubchannel interference. In practical systems, the orthogonality may be imperfect due to, e.g., mismatched oscillators, Doppler frequency shift, and/or timing synchronization errors. In general, the larger the block size N and as such the smaller the subcarrier spacing, the significant the intersubchannel interference due to frequency and timing offset. The intersubchannel interference due to frequency offset is much more severe than those due to timing offset.

Another challenge faced by multicarrier transmission system is potentially very poor channel quality on some subchannels. Recall that the subchannel gains are given by

$$H(i) = \frac{1}{N} \sum_{n=0}^{L-1} h_n e^{-j2\pi ni/N}, \quad i = 0, 1, \ldots, N - 1. \tag{6.63}$$

The channel power gain of some subchannels can be small, leading to low received SNR and high bit error rate on those subchannels. There are several candidate solutions to compensate the performance degradation on certain subchannels. The most popular solution is coding with interleaving over time and frequency. The basic idea is to transmit the coded bits after interleaving over different subcarriers such that the coded bits for the same data bits experience independent fading. The resulting system will enjoy both coding gain and diversity benefit while requiring no channel state information at the transmitter. When the channel state information can be made available at the transmitter, especially for slowly changing channels, adaptive rate/power loading on subchannels is an attractive alternative solution. The basic idea is to vary the power and data rate for subchannels based on their channel gain. With optimal rate/power loading on subcarriers, we can effectively approach the capacity of selective fading channels. We will discuss these two solutions in further details in the following chapters.

6.6 Further readings

Further discussion about other equalization schemes, including decision feedback equalization and maximum likelihood sequence estimation, and their performance can be found in [1, Chapter 10] and [2, Chapter 7]. Reference [1, Chapter 11] discusses the adaptive implementation of equalization technologies. Reference [3] provides a more comprehensive presentation about multicarrier transmission, including its application in various wireless systems. Various PAPR reduction schemes are presented in [4].

Problems

1. Consider a digital transmission system over a wireless channel with transmission power 10 mW. The power of additive noise at the detector input is 10^{-1} mW. Assume that the wireless channel introduces frequency-selective fading with instantaneous tapped-delay channel specified as $L = 5$, $h_0 = 0.1e^{j\pi/5}$, $h_1 = 0.05e^{j7\pi/4}$, $h_2 = 0.08e^{j\pi/6}$, $h_3 = 0.01e^{j3\pi/4}$, and $h_4 = 0.003e^{j\pi/3}$. Determine the SINR at the detector input. What if the noise power is negligible?

2. Because of the multipath effect of wireless channels, the matched filter output at the end of the ith symbol interval is given by

$$r_i = x_i - 0.4x_{i-1} + n_i, \tag{6.64}$$

   where x_i is the ith transmitted symbol, n_i is the AWGN noise with zero mean and variance $N_0/2$.
   (i)   Design an infinite-tap ZF linear equalizer for this channel.
   (ii) Determine the noise variance of the equalized signal with the ZF equalizer designed in (i).
   (iii) Because of the complexity and delay constraints, the ZF equalizer designed in (i) has to be truncated into a two-tap filter. Determine the complex baseband model of the equalized system with the truncated equalizer. Will there be any ISI in the equalized system?

3. A selective fading channel has the following channel vector

$$\mathbf{h} = [0.1e^{j\pi/5}, 0.05e^{j7\pi/4}, 0.08e^{j\pi/6}, 0.01e^{j3\pi/4}, 0.003e^{j\pi/3}]. \tag{6.65}$$

   The normalized noise spectrum density N_0/E_s is 0.1. Determine the tap weights for (i) a 4-tap ZF equalizer and (ii) a 4-tap MMSE equalizer for this channel. What are the impulse responses of the resulting channels after equalization?

4. Multicarrier transmission system with channel bandwidth of 3 MHz is operating over a channel environment with RMS delay spread of 80 µs.
   (i) Determine the minimum number of subcarriers we need to use such that each subchannel will experience frequency flat fading.
   (ii) Assume conventional multicarrier reception with nonoverlapping subchannels. The guard band ratio needs to be 0.2. What is the maximum symbol rate of each subchannel?
   (iii) Compare the total symbol data rate of conventional multicarrier transmission with single-carrier transmission?

5. Show that the minimum spacing between subcarrier frequency to achieve orthogonal subcarriers is equal to $1/T_N$, where T_N is the symbol period of each subcarrier.

6. (i) Calculate the linear convolution of the sequences $x[n] = \{1, 2, 3, 2\}$ and $h[n] = \{1, 2, -1, 0\}$; (ii) Calculate the linear convolution of the sequences $x[n] = \{3, 2, 1, 2, 3, 2\}$ and $h[n] = \{1, 2, -1, 0\}$ and only keep the first six samples; (iii) Calculate the circular convolution of the sequences of $x[n] = \{1, 2, 3, 2\}$ and $h[n] = \{1, 2, -1, 0\}$ using the following formula

$$y[n] = \sum_{k=0}^{N-1} h[k]x[((n-k))_N], \quad n = 0, 1, 2, 3, \tag{6.66}$$

where $((\cdot))_N$ stands for mod N operation. Note that $N = 4$ here. What is your observation when compare the results of (ii) and (iii).

7. Consider the transmission of the following bit stream 01101110001011000 (starting with "01") with QPSK modulation scheme. The following mapping scheme is adopted

Bit pairs	11	01	00	10
Symbol phases	$\pi/4$	$3\pi/4$	$-3\pi/4$	$-\pi/4$

and the symbol amplitude is normalized to 1. The discrete baseband wireless channel is characterized by vector $\{0.5e^{j\pi/3}, 0.2e^{j\pi/5}\}$, i.e., $L = 2$. To address the frequency selectivity, an OFDM transmission scheme with $N = 4$ is applied. The noise effect is neglected here.
   (i) Determine the data symbols, the transmitted channel symbols, and the received channel symbols over eight symbol periods, assuming that no cyclic prefix is added. Apply 4-point DFT, the received channel symbols, four symbols at a time.
   (ii) Now assume two cyclic prefix symbols are added to each four symbol block. Determine the transmitted channel symbols and received channel symbols over 12 symbol periods. Apply 4-point DFT, the received channel symbols, four symbols at a time, after removing the symbols corresponding to cyclic prefix.

(iii)   Calculate $Y[j]$s using the formula

$$Y[j] = H[j]X[j], \quad j = 0, 1, 2, 3, \tag{6.67}$$

where $X[j]$ are data symbols and $H[j]$ are the 4-point DFT of channel vector, and compare it with your result from parts (i) and (ii). Again, this needs to be repeated for every four symbols.

8. Calculate the PAPR of the modulated signal and the transmitted signal in previous problem.

9. Consider an OFDM transmission system with $N = 128$ subcarriers. Of these subcarriers, 114 subcarriers are used for data symbol transmission. The channel bandwidth is 50 MHz. The cyclic prefix in each OFDM block consists of 32 symbols to remove IBI. (i) What is the channel symbol rate of this system? (ii) What is the data symbol rate of this system? (iii) Assume 32-QAM modulation scheme and rate-2/3 coding scheme. Determine the information rate of this system.

Bibliography

[1]   J. G. Proakis, *Digital Communications*, 4th ed. New York, NY: McGraw-Hill, 2001.

[2]   G. L. Stüber, *Principles of Mobile Communications*, 2nd ed. Norwell, MA: Kluwer Academic Publishers, 2000.

[3]   A. R. S. Bahai, B. R. Saltzberg, and M. Ergen, *Multi-Carrier Digital Communications – Theory and Applications of OFDM*, 2nd ed. New York, NY: Springer-Verlag, 2004.

[4]   J. Tellado, *Multicarrier Modulation with Low PAR: Applications to DSL and Wireless*, Boston, MA: Kluwer Academic Publishers, 2000.

Chapter 7
Spread-spectrum transmission

Spread-spectrum transmission refers to the class of transmission technologies that transmission uses much larger bandwidth than that of the modulated signal. The transmitted signal, generated by spreading the spectrum of the modulated signal, will necessarily experience frequency-selective fading, while the direct transmission of modulated signal may experience flat fading. Utilizing more bandwidth than necessary creates several important advantages to spread-spectrum transmission systems. First of all, spread-spectrum transmission is resistant to intentional narrowband jamming. Second, the transmitted signal can hide below the noise floor and make eavesdropping very difficult. These two features of spread-spectrum transmission make it especially suitable for military applications. Combined with a RAKE receiver, spread-spectrum transmission can explore path diversity by combining different multipath components coherently and improve transmission reliability. Spread spectrum also facilitates the sharing of radio spectrum among multiple users in the code domain, leading to the code division multiple access (CDMA) scheme. These two features are exploited in various commercial wireless communication systems. Finally, the wide spectrum of spread-spectrum signal facilitates certain localization applications.

In this chapter, we present the principle of spread-spectrum transmission and demonstrate the basic underlying mechanisms for its desirable features. We first discuss the most popular spectrum spreading method, *direct sequence spread spectrum (DSSS)*. Noting that the desirable features of spread spectrum system rely heavily on the spreading signal design, we then discuss the design of spreading codes. RAKE receiver and multiple user transmission based on DSSS are considered afterwards. We conclude the chapter with a brief discussion on *frequency-hopping spread spectrum (FHSS)*.

7.1 Direct-sequence spread spectrum

Spread-spectrum transmission uses data-independent spreading code to spread the spectrum of modulated signals. With direct-sequence implementation, the spectrum is spread by directly multiplying the modulated signal with a spreading signal generated using a certain spreading code. The structure of the DSSS transmitter is shown in Figure 7.1. While mostly linear modulation schemes can apply, we assume binary

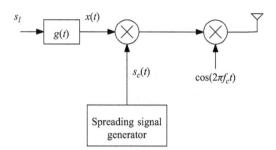

Figure 7.1   Direct-sequence spread spectrum transmitter

phase shift keying (BPSK) with rectangular pulse shape for the sake of presentation clarity. As such, the modulated symbols over each symbol period, denoted by s_l, will take values of $+1$ and -1. The baseband modulated signal over lth symbol period after pulse shaping is given by

$$x(t) = s_l \cdot g(t), \quad 0 \leq t \leq T_s, \tag{7.1}$$

where T_s is the symbol period and $g(t)$ is the rectangular shaping pulse. The modulated signal is then multiplied by the spreading signal $s_c(t)$, which consists of a sequence of short chips, given by

$$s_c(t) = \sum_{n=0}^{G-1} d_n \, g_c(t - nT_p), \quad 0 \leq t \leq T_s, \tag{7.2}$$

where T_p is the chip period, $d_n = \pm 1$ are chip values, and $g_c(t)$ is the chip shaping pulse. Here, the constant G is equal to T_s/T_p and usually referred to as the *processing gain* of the spread spectrum system. Typically, G is chosen to be much greater than 1. Note that the bandwidth of $x(t)$ is proportional to $1/T_s$ and that of $s_c(t)$ to $1/T_p$, which is G times larger than that of $x(t)$. As such, the bandwidth of the transmitted bandpass signal, after carrier modulation is given by

$$x_c(t) = x(t)s_c(t) \cos(2\pi f_c t), \tag{7.3}$$

will be proportional to $1/T_p$, much larger than that of $x(t)$. The spectrum spreading process in time and frequency domain is illustrated in Figure 7.2.

Example: Spread-spectrum transmission

Consider the digital transmission over a wireless channel with RMS (root mean square) delay spread σ_T equal to 0.5 μs. The data symbol rate of the modulated signal is 24 ksps and the roll-off factor of the pulse shape is equal to 0, i.e., rectangular pulse. Will the transmitted signal experience frequency flat or selective fading? What if the modulated signal is spread by a spreading signal with chip rate 1,028 ksps?

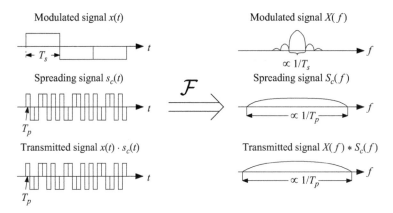

Figure 7.2   Illustration of spectrum spreading in time and frequency domain

Solutions: Since the roll-off factor is equal to 0, we have $T_s B_s = 1$. The bandwidth of the modulated signal is approximately equal to 24 kHz. The coherence bandwidth of the channel is estimated as $B_c = 0.2/\sigma_T = 400$ kHz, which is much greater than B_s. Therefore, the direct transmission of modulated signal will experience frequency flat fading.

   With spread-spectrum transmission, the bandwidth of the transmitted signal is proportional to that of the spreading signal, which is approximately 1.028 MHz and larger than the channel coherence bandwidth of 400 kHz. Therefore, the transmitted signal with spread-spectrum transmission will experience frequency-selective fading.

   The chip period T_p is typically smaller than the RMS delay spread of the wireless channel σ_T. The transmitted signal will experience frequency-selective fading. Assuming that the path delays are integer multiples of chip period T_p, the instantaneous impulse response of the channel can be written as

$$c(\tau) = \sum_{i=0}^{L-1} z_i \delta(\tau - iT_p), \tag{7.4}$$

where z_i is complex gain for ith multipath, $i = 0, 1, \dots, L - 1$. Here, $z_i = 0$ implies that there is no path with delay iT_p. The received signal over such multipath channel is given by

$$r(t) = c(t) * x_c(t) + n(t) = z_i x_c(t - iT_p) + \sum_{j=0, j \neq i}^{L-1} z_j x_c(t - jT_p) + n(t), \tag{7.5}$$

where $n(t)$ is the additive white Gaussian noise. Here, we assume, without loss of generality, that the receiver is interested in the signal copy with delay iT_p. As such, the

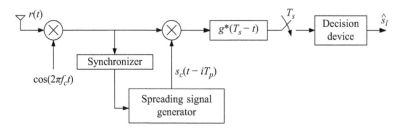

Figure 7.3   Direct-sequence spread spectrum receiver

term $\sum_{j=0,j\neq i}^{L-1} z_j x_c(t-jT_p)$ will act as interference, termed as *multipath interference*. We now examine the effect of DSSS receiver on multipath interference.

The structure of DSSS receiver is shown in Figure 7.3. The received signal is first downconverted to baseband with locally generated carrier. Then, the received signal is multiplied with a locally generated spreading signal, which is synchronized to a particular propagation path, usually the first one with the large enough path power gain. After that, the receiver applies matched filter to perform symbol-by-symbol detection. To elaborate further, let us assume that the receiver is synchronized to the path with delay iT_p and examine the matched filter output corresponding to the signal received over synchronized path. The baseband received signal over the *l*th symbol period corresponding to the signal copy received over the synchronized path is given by

$$z_i x(t-iT_p)s_c(t-iT_p) = z_i s_l\, g(t-iT_p)s_c(t-iT_p). \tag{7.6}$$

The corresponding matched filter output, after multiplying the locally generated spreading signal $s_c(t-iT_p)$, is given by

$$\frac{1}{T_s}\int_{iT_p}^{T_s+iT_p} z_i \cdot s_l\, s_c^2(t-iT_p)dt = z_i \cdot s_l. \tag{7.7}$$

The signal on the synchronized path is perfectly despread for detection.

Meanwhile, the baseband received signal corresponding to those unsynchronized paths is given by

$$\sum_{j=0,j\neq i}^{L-1} z_j x(t-jT_p)s_c(t-jT_p), \tag{7.8}$$

After multiplying the locally generated spreading signal $s_c(t-iT_p)$ and applying the matched filter, we can show that the corresponding sample at the matched filter output is given by

$$\sum_{j=0,j\neq i}^{L-1} \frac{1}{T_s}\int_{iT_p}^{T_s+iT_p} z_j s_l s_c(t-jT_p)s_c(t-iT_p)dt = \sum_{j=0,j\neq i}^{L-1} z_j s_l \rho((i-j)T_p), \tag{7.9}$$

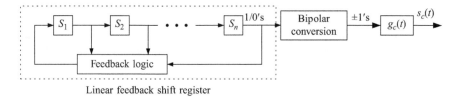

Linear feedback shift register

Figure 7.4   Spreading signal generator based on m-sequence

where $\rho(\tau)$ is autocorrelation function of spreading signal $s_c(t)$, defined at

$$\rho(\tau) = \frac{1}{T_s} \int_0^{T_s} s_c(t) s_c(t - \tau) dt. \tag{7.10}$$

From these results, we can see that multipath interference will be greatly reduced if we can design the spreading signal such that $\rho((i - j)T_p) \cong 0$, for $j \neq i$.

7.1.1   Spreading code design

Recall the spreading signal over a symbol period given in (7.2). The autocorrelation function of the spreading signal evaluated at integer multiples of T_ps can be evaluated as

$$\rho(kT_p) = \frac{1}{G} \sum_{n=1}^{G} d_n d_{n-k}, \tag{7.11}$$

which is dependent upon the spreading sequence $\{d_n\}_{n=1}^{G}$. The spreading signal design problem boils down to the design of proper spreading sequence to achieve the desired autocorrelation function of $s_c(t)$. Ideally, we need an autocorrelation function of the form

$$\rho(\tau) = \begin{cases} 1, & \tau = 0; \\ 0, & \tau \neq 0. \end{cases} \tag{7.12}$$

Unfortunately, only pure noise sequence has such autocorrelation function. Since the receiver needs to independently regenerate the same spreading sequence for despreading process, pure noise sequence cannot be used. Instead, we need spreading sequences that have noise-like properties but can be deterministically generated. Such sequences are called *pseudo-noise (PN)* sequences.

Maximum length sequence, also known as *m*-sequence, is one of the most widely used PN sequences. It also finds application in the synchronization process. *m*-sequences are generated using a linear feedback shift register (LFSR). A spreading signal generator based on *m*-sequence is shown in Figure 7.4. The shift register with linear feedback logic creates a binary sequence of 1s and 0s, which are converted into bipolar format (change 0s to -1s). The spreading signal is obtained after passing the

Table 7.1   Sample feedback logics for shift register

# of Stages	Feedback logics	Output sequence (initial state: all ones)
3	$x_2 \oplus x_3$	11100 10
4	$x_3 \oplus x_4$	11110 00100 11010
5	$x_2 \oplus x_5$	11111 00110 10010 00010 10111 01100 0

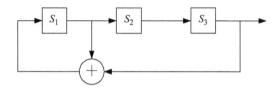

Figure 7.5   Linear feedback shift register with $n = 3$

resulting signal through a chip pulse shaping filter $g_c(t)$. m-sequence refers to the output binary sequence of LFSR with the maximum period. A shift register with n stages will have a maximum of $2^n - 1$ nonzero states. Proper feedback logic is required for the shift register to experience all nonzero states and generate output sequence with period $2^n - 1$. Table 7.1 presents sample feedback logics and the corresponding output m-sequences for different shift register sizes.

m-Sequence has the desired noise-like properties. Specifically, an m-sequence of length $2^n - 1$ has 2^{n-1} ones and $2^{n-1} - 1$ zeros and therefore is quite balanced. Half of the runs (consecutive zeros and ones) in an m-sequence are of length 1, quarter of the runs are of length 2, and so forth. There is one run of length n and another run of length $n - 1$. So, m-sequence has noise-like run length property. Finally, the autocorrelation function of a spreading signal created from an m-sequence of length $2^n - 1$ can be shown to be given by

$$\rho_c(\tau) = \begin{cases} 1 - \frac{|\tau|}{T_p}\left(1 + \frac{1}{2^n-1}\right), & |\tau| \le T_p; \\ -\frac{1}{2^n-1}, & |\tau| > T_p, \end{cases} \tag{7.13}$$

which well approximates the desired autocorrelation function in (7.12) when $2^n - 1$ is large enough. Note that $\rho_c(kT_p) = -1/(2^n - 1)$ for $k \neq 0$. Therefore, the spreading signal designed using m-sequence can attenuate multipath interference by a factor of $1/(2^n - 1)$.

Example: m-Sequence from linear feedback shift register

Consider the LFSR shown in Figure 7.5. Determine its output sequence when the initial states of the registers are 001. Is it an m-sequence? If yes, what is the autocorrelation of the spreading signal generated with this m-sequence when $\tau = 3T_p$?

Solutions: Note that the feedback logic is $x_1 = x_1 \oplus x_3$. The state evolution of the registers and the output bits are

Clock	S_1 S_2 S_3	Output
0	0 0 1	
1	1 0 0	1
2	1 1 0	0
3	1 1 1	0
4	0 1 1	1
5	1 0 1	1
6	0 1 0	1
7	0 0 1	0

$$(7.14)$$

The output sequence 1001110 is an *m*-sequence, as the length is $7 = 2^n - 1$. The autocorrelation of spreading signal generated with this *m*-sequence for $|\tau| > T_p$ is $-1/7$. The balance and run-length properties of the sequence can also be observed.

7.1.2 Effect of DSSS receiver on interference and noise

DSSS receiver may experience certain narrowband interference. Such interference can originate from narrowband systems using the same band or intentional hostile jammer. The received signal on the synchronized path, assuming perfect multipath interference removal, is given by

$$r(t) = z_i x_c(t - iT_p) + n(t) + I(t), \tag{7.15}$$

where $I(t)$ denotes the narrowband interference collected by the receiver. The quality of the received signal is then characterized by the signal-to-interference-plus-noise ratio (SINR), defined as

$$\gamma = \frac{P_s}{P_I + N}, \tag{7.16}$$

where P_s is the signal power, P_I is the interference power, and N is the noise power. Note that the noise power is calculated as the product of signal bandwidth and the noise spectrum density N_0.

Let us first consider the effect of DSSS receiver on the narrowband interference $I(t)$, which is assumed to be of the form

$$I(t) = I_b(t) \cos(2\pi f_c t), \tag{7.17}$$

where $I_b(t)$ is a certain baseband signal of bandwidth B_I. The DSSS receiver will first down convert it to baseband and multiply with a locally generated spreading signal. The resulting signal, given by $I_b(t)s_c(t - iT_p)$, will have a bandwidth of $B_I + B_p$, where B_p is the bandwidth of the spreading signal. As such, the spectrum of the interference signal is spread by the despreading process at the receiver. This process is illustrated

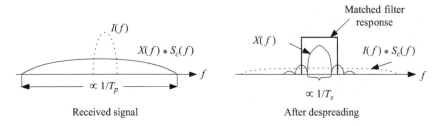

Figure 7.6　Spreading of narrowband interference

in Figure 7.6. The receiver will then apply matched filter detection. The matched filter output over the lth symbol period is given by

$$I_l = \frac{1}{T_s} \int_{iT_p}^{T_s+iT_p} I_b(t)s_c(t - iT_p)dt. \tag{7.18}$$

The matched filter essentially acts as a low-pass filter with bandwidth of $B_s \propto 1/T_s$. As such, the percent of interference signal power that will affect the signal detection is $B_s/(B_I + B_p)$. When $B_p \gg B_I$, $B_s/(B_I + B_p)$ is approximately equal to $1/G$, the inverse of the processing gain of DSSS system. The interference power is reduced by a factor of $1/G$ with spread spectrum receiver.

Similarly, the matched filter output corresponding to the noise signal can be determined as

$$N_l = \frac{1}{T_s} \int_{iT_p}^{T_s+iT_p} n(t)s_c(t - iT_p) \cos(2\pi f_c t)dt. \tag{7.19}$$

It can be shown that the noise component N_l is a Gaussian random variable with zero mean (since $n(t)$ has zero mean) and variance $N_0 B_s$. Essentially, the despreading process will not change the noise property whereas the matched filter again acts as a low-pass filter. Therefore, the SINR at the matched filter output can be approximately determined as

$$\gamma = \frac{P_s}{P_I/G + N_0 B_s}, \tag{7.20}$$

which should be much higher than that at the receive antenna, given by $P_s/(P_I + N_0 B_p)$.

Example: SINR analysis for DSSS receiver

Consider the reception of a DSSS transmission in presence of narrowband interference and noise. The bandwidth of the transmitted signal is 1.25 MHz. The bandwidth of the narrowband interference is 50 kHz and that of the baseband modulated signal is 30 kHz. The received signal power is 10 dBm and the interference power is 15 dBm. The noise spectrum density N_0 is 10^{-8} W/Hz. Determine the SINR

1. at the receive antenna and
2. at the matched filter output.

Solutions:

1. At the receive antenna, the signal power is 10 dBm, or equivalently 0.01 W. The interference power is 0.032 W, whereas the noise power is $N_0 B_p = 0.0125$ W. As such, the SINR is calculated as $\gamma = 0.01/(0.0125 + 0.032) = 0.225 = -6.5$ dB. We can also examine the spectrum density of the received signal, which can be determined as $0.01/1.25 \times 10^{-6} = 8 \times 10^{-9}$ W/Hz. The received signal is buried under the noise floor.

2. At the matched filter output, the signal power is still 0.01 W. The interference power is reduced by a factor of $30/(1{,}250 + 50)$ and becomes 0.74 mW. The noise power becomes $N_0 B_s = 0.4$ mW. Therefore, the SINR after matched filter becomes $\gamma = 10/(0.4 + 0.74) = 8.77 = 9.4$ dB, about 17 dB higher than the SINR at the receive antenna.

7.1.3 RAKE receivers

Basic DSSS receiver synchronizes to a single path and performs detection on the signal received on that path while treating signals from other paths as interference. The synchronized path is typically the first path found during the synchronization process. The multipath signals travel through different propagation mechanisms and carry the same information signal to the receiver. Such potential diversity benefit, usually referred to as *path diversity*, can be fully explored with more complex receiver structures. One intuitive approach is to modify the synchronization process such that the receiver is synchronized to the strongest multipath signal among available ones. Such approach will essentially achieve selection combining (SC) over multipath signals but incur the complexity of estimating and comparing power gains of all resolvable paths.

A more complex receiver will implement multiple branches, with each branch synchronized to a different resolvable path. The resulting receiver structure is shown in Figure 7.7 and widely known as RAKE receiver due to its resemblance to a garden rake. In particular, the jth branch, $j = 1, 2, \ldots, J$, is synchronized to the multipath signal with delay τ_j and as such multiplies the received signal with $s_c(t - \tau_j)$. Typically, the delay difference between branches should be greater than the chip period T_p to extract maximum diversity benefit. The matched filter output of the jth branch over the lth symbol period is given by

$$r_l^{(j)} = z_j s_l + I_m^{(j)} + n_l^{(j)}, \quad j = 1, 2, \ldots, J, \tag{7.21}$$

where $I_m^{(j)}$ denotes the residual multipath interference and $n_l^{(j)}$ is the sample of the additive white Gaussian noise on the jth branch. These outputs are then combined together before being passed to the decision device.

Various combining techniques can apply, whereas maximum ratio combining (MRC) is the most common in practice. Specifically, with properly designed spreading code, the residual multipath interference $I_m^{(j)}$ can be treated as additional noise.

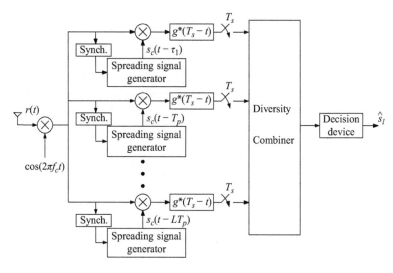

Figure 7.7   Structure of RAKE receiver

With MRC, the matched filter output $r_l^{(i)}$ will be weighted by z_i^* and then summed together, leading to the decision statistics given by

$$\hat{s}_l = \sum_{j=1}^{J} |z_j|^2 s_l + \sum_{j=1}^{J} z_j^* \tilde{n}_l^{(j)}, \qquad (7.22)$$

where $\tilde{n}_l^{(j)} = I_m^{(j)} + n_l^{(j)}$. The receiver can achieve similar performance as an L-branch MRC diversity receiver when $\tilde{n}_l^{(j)}$'s are independent and identically distributed across different branches. Apparently, increasing the number of branches L leads to better performance, but at the cost of higher receiver complexity.

Example: Combined SNR of RAKE receiver

The frequency-selective fading channel experienced by a spread-spectrum transmission system has the following instantaneous realization

$$c(\tau) = 0.2e^{j3\pi/5}\delta(\tau) + 0.41e^{j\pi/3}\delta(\tau - T_p) - 0.32e^{j\pi/3}\delta(\tau - 3T_p). \quad (7.23)$$

The transmitted symbol energy is 10^{-6} J and the noise spectrum density is 10^{-7} W/Hz. Assume that an ideal spreading code is applied, and as such, multipath interference can be neglected. The receiver implements a dual-branch RAKE combiner. Determine the signal-to-noise ratio (SNR) of the decision statistics when

1. receiver branches are synchronized to paths with smaller delays and
2. receiver branches are synchronized to the strongest paths.

Solutions:

1. The SNR of the decision statistics is given, without multipath interference, by

$$\gamma = \sum_{j=1}^{J} |z_j|^2 \frac{E_s}{N_0}. \tag{7.24}$$

When the receiver branches are synchronized to paths with smaller delays, we have $z_1 = 0.2$ and $z_2 = 0.41$. The SNR is determined as

$$\gamma = \left(0.2^2 + 0.41^2\right) \frac{10^{-6}}{10^{-8}} = 20.8 = 13.2 \text{ dB}. \tag{7.25}$$

2. When the receiver branches are synchronized to the strongest paths, we have $z_1 = 0.41$ and $z_2 = 0.32$. The resulting SNR is

$$\gamma = \left(0.41^2 + 0.32^2\right) \frac{10^{-6}}{10^{-8}} = 27.1 = 14.3 \text{ dB}. \tag{7.26}$$

Note that the SNR gain by synchronizing to the strongest paths comes at the additional complexity of ranking all available paths according to the power gains.

7.2 Multiple access with CDMA

DSSS transmission also facilitates the multiple access scheme: CDMA. With CDMA, multiple users can transmit/receive over the same spectrum at the same time, but using different spreading codes. Let us consider the downlink transmission from a base station to K different mobile users. $x_k(t)$ denotes the modulated signal targeted to user k, $k = 1, 2, \ldots, K$. Before transmitting $x_k(t)$, together with other user signal, the base station multiplies a unique spreading signal $s_{c,k}(t)$ to it, as illustrated in Figure 7.8. The resulting transmitted signal from the base station, after carrier modulation, is given by

$$\tilde{x}_c(t) = \sum_{k=1}^{K} x_k(t) s_{c,k}(t) \cos(2\pi f_c t). \tag{7.27}$$

Assuming a frequency-selective fading environment, the received signal at user k is given by

$$r_k(t) = z_{i,k} \tilde{x}_c(t - iT_p) + \sum_{m=0, m \neq i}^{L-1} z_{m,k} \tilde{x}_c(t - mT_p) + n_k(t), \tag{7.28}$$

where $z_{i,k}$ is the complex gain of the path with delay iT_p for user k and $n_k(t)$ is the additive white Gaussian noise at user k. Note that $r_k(t)$ contains transmitted signals targeted at other users, which act as interference to the detection of $x_k(t)$. Such interference is usually referred to as *multiuser interference*.

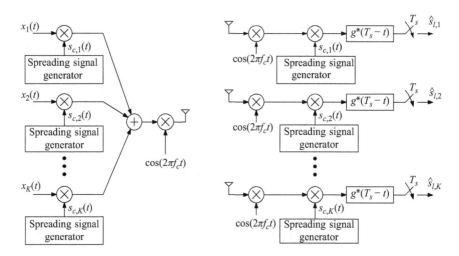

Figure 7.8   Multiuser transmission based on CDMA

User k will apply the standard DSSS receiver to process $r_k(t)$. Specifically, the received signal is first down converted to baseband with carrier multiplication. Without loss of generality, let us focus on the received baseband signal over the synchronized path, which can be rewritten as

$$z_{i,k}x_k(t - iT_p)s_{c,k}(t - iT_p) + z_{i,k} \sum_{j=1,j\neq k}^{K} x_j(t - iT_p)s_{c,j}(t - iT_p). \tag{7.29}$$

The second term is the so-called multiuser interference, which should be eliminated or reduced as much as possible before performing detection. The receiver multiplies user ks spreading signal with the corresponding delay, $s_{c,k}(t - iT_p)$, to the received signal and then apply matched filter detection. The matched filter output over the lth symbol period can be shown to be given by

$$\hat{s}_{l,k} = z_{i,k}s_{l,k} + \frac{1}{T_s} \int_{iT_p}^{T_s+iT_p} z_{i,k} \sum_{j=1,j\neq k}^{K} x_j(t - iT_p)s_{c,j}(t - iT_p)s_{c,k}(t - iT_p)dt$$

$$= z_{i,k}s_{l,k} + \sum_{j=1,j\neq k}^{K} z_{i,k}s_{l,j}\rho_{jk}(0), \tag{7.30}$$

where $\rho_{jk}(\cdot)$ denotes the crosscorrelation function between the spreading signals of user k and user j, defined as

$$\rho_{jk}(\tau) = \frac{1}{T_s} \int_0^{T_s} s_{c,j}(t)s_{c,k}(t - \tau)dt. \tag{7.31}$$

Therefore, if the spreading signals used by different users have small crosscorrelation, then the effect of multiuser interference can be greatly attenuated.

7.2.1   Spreading codes for multiple access

The multiuser interference rejection capability of DSSS transmission depends on the crosscorrelation property of the spreading signals. Since the user signals for downlink transmission are synchronized, we only need $\rho_{jk}(\tau)$ to be small for $\tau = 0$ to attenuate the multiuser interference over the synchronized path. Meanwhile, to attenuate the multiuser interference on unsynchronized paths, we also need $\rho_{jk}(\tau)$ for $\tau = mT_P$ to be small. Furthermore, the uplink transmission is typically asynchronous as users are geographically distributed. Transmitted signal from different users will experience different propagation delay before reaching the base station. In this case, the multiuser interference attenuation would require $\rho_{jk}(\tau)$ to be small over a continuum value range of τ.

The crosscorrelation property of the spreading signals again depends on the codes used to create the spreading signal. In particular, the spreading signal of user k over a symbol period can be written as

$$s_{c,k}(t) = \sum_{n=1}^{G} d_{n,k}\, g_c(t - nT_p), \quad 0 \le t \le T_s, \tag{7.32}$$

where $\{d_{n,k}\}_{n=1}^{G}$ is the spreading code for user k. Then, the crosscorrelation function between two spreading signals can be calculated as

$$\rho_{jk}(\tau) = \frac{1}{T_s} \int_0^{T_s} \left(\sum_{n=1}^{G} d_{n,j}\, g_c(t - nT_p) \right) \left(\sum_{n=1}^{G} d_{n,k}\, g_c(t - \tau - nT_p) \right) dt. \tag{7.33}$$

If the relative delay $\tau = mT_p$, then it can be shown that $\rho_{jk}(\tau)$ is related to the spreading codes as

$$\rho_{jk}(\tau) = \frac{1}{G} \sum_{n=1}^{G} d_{n,j} d_{((n+m))_G,k}, \tag{7.34}$$

where $((\cdot))_G$ denotes the modulo-G operation. We now present sample spreading codes that have good crosscorrelation properties.

Walsh–Hadamard codes are perfectly orthogonal when synchronized. Therefore, the spreading signal designed with such codes will lead to $\rho_{jk}(0) = 0$. The Walsh–Hadamard codes of length 2^N are rows of matrix H_{2^N}, which is recursively defined as

$$H_{2^N} = \begin{bmatrix} H_{2^{N-1}} & H_{2^{N-1}} \\ H_{2^{N-1}} & -H_{2^{N-1}} \end{bmatrix} \text{ with } H_2 = \begin{bmatrix} 1 & 1 \\ 1 & -1 \end{bmatrix}. \tag{7.35}$$

For example, H_4 are given by

$$H_4 = H_{2^2} = \begin{bmatrix} H_2 & H_2 \\ H_2 & -H_2 \end{bmatrix} = \begin{bmatrix} 1 & 1 & 1 & 1 \\ 1 & -1 & 1 & -1 \\ 1 & 1 & -1 & -1 \\ 1 & -1 & -1 & 1 \end{bmatrix}. \tag{7.36}$$

There are a total of 2^N distinct Walsh–Hadamard codes of length 2^N. While orthogonal when synchronized, the orthogonality between these codes will be destroyed without synchronization. In fact, the shifted version of one Walsh–Hadamard code may become another Walsh–Hadamard code (see the third and fourth row of H_4 above). As such, Walsh–Hadamard codes are not suitable for CDMA uplink transmission where user synchronization is very challenging.

Gold codes are another type of spreading codes that can be used in CDMA systems. Noting that achieving perfect orthogonality among asynchronous users is extremely difficult in real-world environment, Gold codes are nonorthogonal codes with bounded crosscorrelation functions. These codes are generated by adding two preferred pairs of m-sequences in modulo-2 fashion. These preferred pairs of m-sequences are chosen such that the crosscorrelation of two Gold codes over a code period takes only three possible values as

$$\rho_{jk}(\tau) = \begin{cases} -\frac{1}{2^n - 1}; \\ -\frac{t(n)}{2^n - 1}; \\ \frac{t(n) - 2}{2^n - 1}, \end{cases} \tag{7.37}$$

where n is the size of the linear shift register used to produce m-sequences and

$$t(n) = \begin{cases} 2^{(n+1)/2} + 1, & n \text{ odd;} \\ 2^{(n+2)/2} + 1, & n \text{ even.} \end{cases} \tag{7.38}$$

It is interesting to note that the autocorrelation of Gold codes also takes these three possible values. As such, Gold codes possess a certain multipath interference mitigation capability, although not as good as that of original m-sequences.

7.2.2 Performance analysis

The detection performance of CDMA receivers depends on the multiuser interference rejection capability of the spreading signals. Because of the usage of nonorthogonal codes and/or synchronization difficulty, there always exists residual multiuser interference. In fact, the performance of CDMA receivers is typically interference limited. Their exact performance analysis would demand the specification of spreading codes and symbol transmission schemes. To appreciate the involved design tradeoffs without being deluged with tedious details, we adopt a simplified approach by assuming that the effective multiuser interference from each user is well approximated by independent identical Gaussian process with power P_s/G, where G is the processing gain of the system. As such, the signal-to-interference ratio (SIR) of the received signal at the kth user in a K-user downlink transmission system (referring to (7.30)) is given by

$$\text{SIR} = \frac{G}{K - 1}. \tag{7.39}$$

Note that the effect of channel power gain is common to both desired signal and multiuser inference and therefore canceled. If the noise power needs to be taken into account, we can calculate the resulting SINR as

$$\text{SINR} = \left(\frac{K-1}{G} + \frac{N_0}{|z_{i,k}|^2 E_s} \right)^{-1}, \tag{7.40}$$

where $z_{i,k}$ is the gain of the synchronized path for user k. Essentially, the SINR is calculated as the inverse of the sum of interference power and noise power, both normalized by received signal power. The error rate performance of CDMA receiver can then be evaluated based on the adopted modulation schemes.

Example: User capacity of CDMA system

Consider a CDMA-based wireless communication system in a multiuser interference limited environment. The system adopts BPSK modulation scheme and the processing gain is equal to 60. Determine the number of users that the system can accommodate such that the experienced bit error rate (BER) of each user is at most 10^{-4}. What if the error rate requirement becomes 10^{-3}?

Solutions: Recall the BER of BPSK modulation scheme is $P_b = Q(\sqrt{2\gamma})$. To satisfy the BER requirement of 10^{-4}, we need $\sqrt{2\gamma}$ to be at least 3.7. Plugging in the SIR expression for γ, we have

$$\sqrt{2\frac{G}{K-1}} \geq 3.7, \tag{7.41}$$

from which the number of users K can be shown to be less than 9.8. As such, at most 9 users can be supported simultaneously.

　　If the error rate requirement is changed to 10^{-3}, we need $\sqrt{2\gamma}$ to be at least 3.1. The maximum number of users that can be supported is determined to be 13. We can notice that unlike FDMA and TDMA systems, CDMA systems have soft capacity.

7.3   Frequency-hopping spread spectrum

FHSS implementation spreads the spectrum of modulated signal by changing its carrier frequency over a wide bandwidth according the spreading signal $s_c(t)$. The frequency-hopping process is illustrated in Figure 7.9, where the carrier frequency is changed to a new value for every chip period T_p. If there are N carrier frequencies available for hopping and the modulated signal bandwidth is B_s, then the bandwidth of frequency-hopping transmission system is NB_s. When T_p is larger than the symbol period T_s, i.e., $T_p = kT_s$, where k is an integer, the hopping is deemed relatively slow, leading to the so-called slow frequency hopping (SFH) system. On the other hand, when $T_p = T_s/k$, i.e., the carrier frequency changes multiple times per symbol, we have fast frequency hopping (FFH). FFH systems can exploit frequency diversity

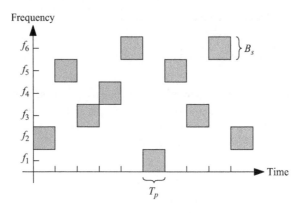

Figure 7.9   Frequency-hopping transmission

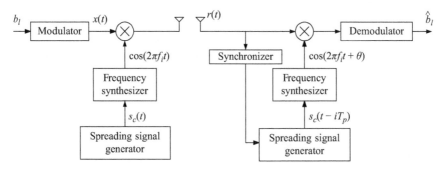

Figure 7.10   Structure of frequency-hopping transmitter and receiver

within each symbol period, whereas SFH systems need to rely on coding/interleaving schemes to extract frequency diversity.

The structure of frequency-hopping transmitter and receiver is shown in Figure 7.10. The spreading signal $s_c(t)$ is fed into the frequency synthesizer to control the carrier frequency. Unlike DSSS implementation, the values of $s_c(t)$ over each chip period T_p will be the index of carrier frequencies. Due to the difficulty of maintaining phase coherence, noncoherent and differentially coherent modulation schemes are typically used in frequency-hopping systems. The receiver will first synchronize the local spreading signal generator. The synchronized spreading signal is then used to control the frequency synthesizer, which generates carrier for down conversion.

We now examine the anti-jamming capability of FHSS system in more details. Let us consider an uncoded SFH system where the modulated signal uniformly hops over N carrier frequencies. An narrowband jammer generates constant interference power of P_I over M out of these N carrier frequencies. Therefore, the probability that the transmission is affected by the jamming signal over a chip period is M/N.

Conditioning on whether the transmitted symbol is affected by the jamming interference or not, the average error rate of frequency-hopping transmission system is calculated as

$$\overline{P}_E = \Pr[\text{without interference}]\overline{P}_{E|NI} + \Pr[\text{with interference}]\overline{P}_{E|WI}$$
$$= (1 - M/N)\overline{P}_{E|NI} + (M/N)\overline{P}_{E|WI}, \tag{7.42}$$

where $\overline{P}_{E|NI}$ is the average error rate over carriers not affected by jamming interference and $\overline{P}_{E|WI}$ is the average error rate over carriers with interference. Let us assume that the average error rate of the chosen modulation scheme with the consideration of fading effect can be calculated as $\overline{P}_E(\overline{\gamma}_s)$, where $\overline{\gamma}_s$ is the average receive SNR or SINR. For symbols not affected by jamming, the received SNR is given by P_s/N_0B_s, where P_s is the average received signal power. When interference is present, we can treat the interference as additional noise component with power P_I and determine the average received SINR as

$$\overline{\gamma}_s = \frac{P_s}{N_0B_s + P_I}. \tag{7.43}$$

Finally, the average error rate for the frequency-hopping transmission in presence of jamming interference can be calculated as

$$\overline{P}_E = (1 - M/N)\overline{P}_E\left(\frac{P_s}{N_0B_s}\right) + (M/N)\overline{P}_E\left(\frac{P_s}{N_0B_s + P_I}\right). \tag{7.44}$$

Example: Error performance of frequency-hopping transmission

A baseband modulated signal of bandwidth 25 kHz is transmitted in a frequency-hopping fashion over a spectrum of 1.25 MHz. The transmission is suffering a narrowband jamming interference of bandwidth 50 kHz (affecting the transmission over two carriers). The transmission over each carrier experience independent and identical Rayleigh fading with average received signal power is 20 dBm. The total received interference power is 10 dBm. The noise spectrum density N_0 is 5×10^{-8} W/Hz. Assuming that the system adopts a binary differential phase shift keying (DPSK) modulation scheme, whose average error rate is related to the average received SNR/SINR $\overline{\gamma}$ as $1/(2(1 + \overline{\gamma}))$, determine the average bit error probability of the transmission system.

Solutions: Given the channel bandwidth, the number of carrier frequencies available for hoping is 50. We have $M = 2$ and $N = 50$. The average error probability is calculated as

$$P_E = \left(1 - \frac{2}{50}\right)\frac{1}{2(1 + P_s/N_0B_s)} + \frac{2}{50}\frac{1}{2(1 + P_s/(N_0B_s + P_I))} \tag{7.45}$$

$$= \frac{24}{25}\frac{1}{2(1 + 100/1.25)} + \frac{1}{25}\frac{1}{2(1 + 100/(1.25 + 10))}. \tag{7.46}$$

We can determine the average error rate as $P_E = 0.0062 \times 24/25 + 0.05 \times 1/25 = 0.0079$. Note that the error probability without interference is 0.0062. The effect of jamming interference slightly increases the error rate.

7.4   Further readings

This chapter serves as an introduction of spread-spectrum transmission. For more comprehensive treatment on spread-spectrum transmission, the readers are referred to many excellent books [1–3]. Reference [1, Chapter 13] presents the more detailed analysis of spread-spectrum transmission over AWGN channels. Reference [2, Chapter 13] provides more details on the performance in multiuser environment. The spreading sequence design and performance analysis is available in [3, Chapter 9]. Reference [4] discusses various advantages and challenges of CDMA systems.

Problems

1.   A DSSS transmission system is using an *m*-sequence of length 63 to spread the modulated signal. The wireless channel introduces selective fading with instantaneous impulse response

$$c(\tau) = 0.2e^{j3\pi/5}\delta(\tau) + 0.41e^{j\pi/3}\delta(\tau - T_p) - 0.32e^{j\pi/3}\delta(\tau - 3T_p). \qquad (7.47)$$

Assuming that the DSSS receiver is synchronized to the first path and the noise effect is negligible, determine matched filter output over the *i*th symbol period.

2.   Consider the LFSR shown in Figure 7.11 with initial state 0001.

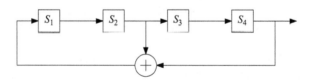

Figure 7.11   Linear feedback shift register with n = 4

i.   What is the period of the output sequence, N?

ii.   How many 1s and 0s are there in one period?

iii.   Determine the value of the *normalized* autocorrelation of the output sequence,

$$\rho(k) = \frac{1}{N}\sum_{i=1}^{N} d_i d_{i+k}, \qquad (7.48)$$

for $k = 1, 2,$ and 3 cases.

3. Assume that in the design of a CDMA system, we need a PN sequence with the following two properties:
   i.   The period of the sequence $N \geq 1,000$;
   ii. The absolute value of the *normalized* autocorrelation function $\mathcal{R}(k) = \frac{1}{N} \sum_{i=0}^{N-1} p_i p_{i-k}$ for $k = 1, 2, \ldots, N-1$ is less than 0.0005.
   Determine the minimum length, i.e., the smallest value of n, of the LFSR that may generate a PN sequence with both properties.

4. Consider the reception of a DSSS transmission in presence of a narrowband interference and noise. The bandwidth of the transmitted signal is 5 MHz. The bandwidth of the narrowband interference is 100 kHz and that of the baseband modulated signal is 30 kHz. The received signal power is 10 dBm and the interference power is 15 dBm. The noise spectrum density N_0 is 10^{-4} mW/Hz. Determine the SINR (i) at the receive antenna and (ii) at the matched filter output.

5. Consider a CDMA system employing BPSK modulation. Assume that there are K active users that are communicating with the base station over the same frequency band at the same time while using different Walsh–Hadamard spreading sequences. Considering the effect of both AWGN noise and interference from other users, the received signal for the first user is given by

$$r(t) = \sum_{i=1}^{K} s_i(t) + n(t), \tag{7.49}$$

where $s_i(t)$ is the signal for the ith user and $n(t)$ is the AWGN noise. Derive the decision statistics for the ith symbol of the first user in presence of both AWGN noise and interference from other users.

6. A CDMA system is employing BPSK with DSSS modulation. The chip rate of each carrier is 1,248 kbps. Assume that the data rate of each active user is 19.2 kbps.
   i.   Considering the effect of multiple access interference (MAI) only, determine the maximum number of simultaneous users that one carrier can support such that the bit error rate experienced by each user is no more than 10^{-3}.
   ii. Now consider the effect of both MAI and noise while assuming that the received signal to noise power ratio is 10 dB. How many users can one carrier support with the same error rate of 10^{-3} for each user now.

7. A cellular telephone service provider deploys a CDMA system to provide full duplex voice communication service to the city of Victoria. The total allocated bandwidth in a cell is 25 MHz. Half of these bandwidth will be used for uplink transmission and the other half for downlink, i.e., frequency division duplexing (FDD). In order to support the chip rate of 1.248 Mbps, each carrier requires 1.25 MHz bandwidth. Assume that each user generates the same data rate of 9.6 kbps and ignore the effect of thermal noise and interference from other cells. Determine the maximum number of users that the cell can support if the minimum required SNR for each user is 10 dB.

8. A baseband modulated signal of bandwidth 25 kHz is transmitted in a frequency-hopping fashion over a spectrum of 5 MHz. The transmission is suffering a narrowband jamming interference of bandwidth 100 kHz (affecting the transmission over four carriers). The transmission over each carrier experience independent and identical Rayleigh fading with average received signal power is 15 dBm. The total received interference power is 20 dBm. The noise spectrum density N_0 is 10^{-8} W/Hz. Assuming that the system adopts BPSK modulation scheme, determine the average bit error probability of the transmission system.

Bibliography

[1] J. G. Proakis, *Digital Communications*, 4th ed. New York, NY: McGraw-Hill, 2001.
[2] A. Goldsmith, *Wireless Communications*. New York, NY: Cambridge University Press, 2005.
[3] G. L. Stüber, *Principles of Mobile Communications*, 2nd ed. Norwell, MA: Kluwer Academic Publishers, 2000.
[4] S. Haykin and M. Hoher, *Modern Wireless Communications*. Upper Saddle River, NJ: Prentice Hall, 2005.

Chapter 8
Channel capacity and coding

Channel capacity characterizes the maximum transmission rate that a channel can support for error-free information delivery. As the performance upper limit for arbitrary transmission system, channel capacity provides valuable guidances for real-world transceiver design. The mathematical theory of channel capacity was established by Claude Shannon in the late 1940s, through his coding theorems. The advanced modulation and coding schemes developed afterward validate Shannon's pioneering vision. Error-control coding serves an effective capacity achieving solution. This chapter studies capacity and coding for wireless fading channels. We first discuss the capacity definition and sample error-control coding schemes for additive white Gaussian noise (AWGN) channels. We then present the commonly used capacity definition for both flat and selective fading channels. In addition to the ergodic capacity and capacity with outage, we also introduce and derive the optimal power and rate adaptation (OPRA) capacity. We conclude the chapter with the discussion of interleaving technique, which is widely used in wireless systems to mitigate the effect of deep fade on coded transmission.

8.1 Capacity of AWGN channels

In general, the capacity of an AWGN channel is given by the Shannon's formula (also known as Shannon–Hartley law) as

$$C = B_s \log_2\left(1 + \frac{P}{N}\right) \text{ bps,} \tag{8.1}$$

where B_s is the channel bandwidth and P/N is the signal-to-noise ratio (SNR) of the received signal. Shannon's coding theorem establishes that when the information data rate R_b is less than or equal to C, there exists a coding scheme that can achieve error-free transmission. Apparently, the capacity for nonzero bandwidth grows to infinity when SNR goes to infinity. On the other hand, we need to maintain a minimum SNR value for the capacity to scale with channel bandwidth B_s.

Let us consider a discrete-time AWGN channel with the received symbol over the ith symbol period given by

$$r_i = s_i + n_i, \tag{8.2}$$

where s_i is the transmitted symbol with bit energy E_b and n_i is the sample of the white Gaussian noise with noise spectral density of $N_0/2$. The transmit power P is equal to $E_b R_b$, where $R_b = 1/T_b$ is the information bit rate. The noise power N is equal to $B_s N_0$. The Shannon's capacity formula specializes for AWGN channel to

$$C = B_s \log_2 \left(1 + \frac{E_b}{N_0} \frac{R_b}{B_s} \right). \tag{8.3}$$

The information data rate R_b should be less than or equal to C for error-free transmission to be possible. Therefore, we arrive at the following inequality, involving the ratio of capacity over signal bandwidth

$$\frac{C}{B_s} \leq \log_2 \left(1 + \frac{E_b}{N_0} \frac{C}{B_s} \right), \tag{8.4}$$

where equality holds for ideal system with $R_b = C$. Solving for E_b/N_0, we can determine the minimum required E_b/N_0 value for error-free transmission as

$$\frac{E_b}{N_0} \geq \frac{B_s}{C} \left(2^{\frac{C}{B_s}} - 1 \right). \tag{8.5}$$

We can see that the SNR E_b/N_0 needs to be greater than $\frac{B_s}{C}(2^{\frac{C}{B_s}} - 1)$ for a certain ratio of C/B_s for C to scale with B_s. If C is much larger than B_s, then the required E_b/N_0 value is very large. On the other hand, if B_s is much larger than C, such than C/B_s approaches 0, the minimum value of E_b/N_0 can be shown to be

$$\lim_{C/B_s \to 0} \frac{B_s}{C} (2^{\frac{C}{B_s}} - 1) = \ln 2 = -1.6 \text{ dB}. \tag{8.6}$$

As such, even when the channel bandwidth is extremely large, we still require a minimum E_b/N_0 value of -1.59 dB to achieve error-free transmission at the rate of $R_b \leq C$. Figure 8.1 illustrates the minimum required E_b/N_0 value for different C/B_s ratio. It is important to note that for a given C/B_s value, error-free transmission is possible only if the received SNR E_b/N_0 is larger than or equal to the corresponding minimum value and the information rate R_b is less than or equal to C. A transmission system can be brought into the error-free possible region by (i) increasing the E_b/N_0 value and/or (ii) reducing the C/B_s ratio, i.e., decreasing the transmission rate R_s or increasing the bandwidth B_s.

Example: Shannon capacity of AWGN channel

Consider an AWGN channel with bandwidth 30 kHz. The E_b/N_0 value of the transmission is 10 dB.

1. What is the maximum possible information rate that the channel can support for error-free transmission?
2. If the required information rate is 300 kbps, what modifications can be made such that error-free transmission is still possible?

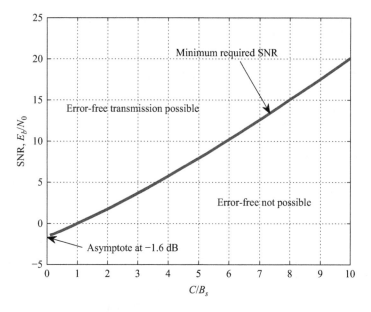

Figure 8.1   Minimum required SNR for given C/B_s ratio

Solutions:

1. For the given E_b/N_0 value of 10 dB, the maximum C/B_s ratio that can be supported for error-free transmission is 5.91. Therefore, the maximum information rate that can be supported is $30 \times 5.91 = 177.3$ kbps.

2. If $R_s = 300$ kbps, the minimum C/B_s ratio will be 10. To achieve error-free transmission, we can increase either the channel bandwidth or the E_b/N_0 value. The minimum channel bandwidth required for the same E_b/N_0 value of 10 dB is $300/5.91 = 50.76$ kHz. The minimum required E_b/N_0 value for the same channel bandwidth will be $(2^{10} - 1)/10 = 20.1$ dB.

Now that the upper limit of the transmission rate for error-free transmission over AWGN channel has been established by Shannon's theorem, the challenge becomes how to approach this performance limit with effective transceiver design. Essentially, we need to devise a system configuration that can support an information rate close to the capacity with arbitrarily small error rate. One approach is to use orthogonal signaling together with correlation-based receiver structure. When the number of orthogonal signals used goes to infinity, the system can achieve the optimal performance, but at the cost of infinite channel bandwidth. Another approach is to use efficient channel coding schemes. The basic idea of channel coding is to introduce controlled redundancy into the transmitted signal to help correct and detect channel-induced errors. Numerous coding schemes, including block codes, convolution codes, concatenated codes, turbo codes, and low-density parity-check (LDPC) codes, have been proposed and extensively studied in the literature for AWGN channels. It has been shown that certain turbo-like coding scheme can perform very close to the capacity limit of Shannon's theorem with reasonable complexity.

8.2 Channel coding for AWGN channels

In this section, we review basic channel coding schemes for AWGN channels. The discussion will concentrate on linear binary codes in this brief introduction.

8.2.1 Block codes

Let us assume that the information source generates a binary data stream of rate R_b bits/s. With an (n, k) block code, for each block of k information bits, the block encoder will produce a codeword of n bits. Here, $n - k$ redundant bits are added to increase the dissimilarity between resulting codewords. The most common measure for dissimilarity between binary sequences is the *Hamming distance*, which is defined as the number of positions that two codewords differ. Note that the minimum Hamming distance between two uncoded information bit sequences of length k is equal to 1. If the minimum distance between the codewords is 3, then the correct codeword can be determined at the receiver based on maximum likelihood decoding rule, when the channel only introduces a single bit error within each codeword.

The simplest block code that achieves such error correction capability is the (3, 1) repetition code. The encoder will generate codeword "111" for each information bit "1" and codeword "000" for each information bit "0." If the channel only introduces one bit error within each codeword, the possible received bit sequence for codeword "111" will be "111," "110," "101," and "011," all of which will lead to the correct decoding of information bit "1" based on the maximum likelihood rule. While enjoying great error-correcting capability, repetition codes suffer the disadvantage of low *code rate*. The code rate of a coding scheme is defined as the number of the information bits per coded bit, usually denoted by R_c. As such, the coding rate of (n, k) block code is equal to k/n. For example, (3, 1) repetition code has a coding rate of $1/3$. The coded transmission will require three times the bandwidth of the uncoded transmission for the same information data stream. As such, it is desirable to design codes with good error-correction capability as well as high coding rate. Unfortunately, these desired properties are generally conflicting with each other.

Example: Block code

Let us consider a block code that generates codewords of length six for every two information bits in the following fashion.

$$00 \Longrightarrow 000101; \tag{8.7}$$

$$01 \Longrightarrow 111101;$$

$$11 \Longrightarrow 011110;$$

$$10 \Longrightarrow 100011.$$

1. Determine the minimum distance between codewords.
2. If 110101 is received, what is the most likely transmitted information bit pair?

Solutions:

1. After checking the Hamming distance between all codeword pairs, the minimum distance can be determined to be 3. So this (6, 2) block code can correct single bit errors.
2. After checking the distance between the received bit sequence with all codewords, we find the codeword 111101 is the most similar to the received sequence, with distance equal to 1. Based on the maximum likelihood rule, we claim that codeword 111101 was transmitted to carry information bit pair 01.

Practical communication systems also prefer coding scheme with certain structure that allows for easy implementation of encoding and decoding process. Without proper structure, as illustrated in previous example, the encoding for (n, k) block code would entail a look-up table of size 2^k and the decoding involve the calculation of 2^k Hamming distances, which requires excessive memory and long processing time. *Parity-check codes* are a class of block codes with certain desirable structure. The encoding and decoding process of parity-check codes uses linear operation with generator matrix and parity-check matrix, respectively. For (n, k) parity-check codes, the generator matrix $\mathbf{G}$ is an $n \times k$ binary matrix. Let $\mathbf{d}$ denote the vector of k information bits. The codeword vector $\mathbf{c}$ of length n is calculated through matrix multiplication as

$$\mathbf{c} = \mathbf{G} \times \mathbf{d}. \tag{8.8}$$

For systematic codes, where the first k bits of the codeword are the information bits, the matrix G has the form of

$$\mathbf{G} = \begin{bmatrix} 1 & 0 & \cdots & 0 \\ 0 & 1 & \cdots & 0 \\ \vdots & \vdots & \ddots & \vdots \\ 0 & 0 & \cdots & 1 \\ h_{11} & h_{12} & \cdots & h_{1k} \\ \vdots & \vdots & \ddots & \vdots \\ h_{r1} & h_{r2} & \cdots & h_{rk} \end{bmatrix} = \begin{bmatrix} \mathbf{I}_k \\ \mathbf{H}_p \end{bmatrix}, \tag{8.9}$$

where $\mathbf{I}_k$ denotes $k \times k$ identity matrix and $\mathbf{H}_p$ is used to calculate the parity-check bits in the codewords.

The decoder multiplies the received codeword $\mathbf{r}$ with parity-check matrix $\mathbf{H}$, which for systematic codes has the form of

$$\mathbf{H} = \begin{bmatrix} h_{11} & h_{12} & \cdots & h_{1k} & 1 & 0 & \cdots & 0 \\ h_{21} & h_{22} & \cdots & h_{2k} & 0 & 1 & \cdots & 0 \\ \vdots & \vdots & \ddots & \vdots & \vdots & \vdots & \ddots & \vdots \\ h_{r1} & h_{r2} & \cdots & h_{rk} & 0 & 0 & \cdots & 1 \end{bmatrix} = [\mathbf{H}_p \ \mathbf{I}_k]. \tag{8.10}$$

If $\mathbf{Hr} = 0$, then $\mathbf{r}$ is a valid codeword. Assuming small probability of bit error, $\mathbf{r}$ will be claimed as the transmitted codeword and the first k bits of $\mathbf{r}$ are detected as the transmitted information bits. It is possible that the received valid codeword is not the transmitted one due to many errors during transmission, leading to undetected errors. If $\mathbf{Hr} \neq 0$, then there will be transmission error. If the channel introduces a single error in each codeword, then the decoder can use multiplication result $H\mathbf{r}$ to correct the single error. Specifically, noting that the received codeword $\mathbf{r} = \mathbf{c} \oplus \mathbf{e}$, where $\mathbf{e}$ is the error vector with only one nonzero entry, we have

$$\mathbf{Hr} = \mathbf{Hc} \oplus \mathbf{He} = \mathbf{He}, \tag{8.11}$$

which is known as the syndrome, denoted by $\mathbf{s}$. Apparently, $\mathbf{s}$ will be a column of H, and the column index indicates the position of transmission error.

Example: Parity-check code

A $(6, 3)$ parity-check code has the following generator matrix

$$G = \begin{bmatrix} 1 & 0 & 0 \\ 0 & 1 & 0 \\ 0 & 0 & 1 \\ 1 & 1 & 0 \\ 0 & 1 & 1 \\ 1 & 0 & 1 \end{bmatrix}. \tag{8.12}$$

Assume that bit sequence 111011 is received, determine if transmission error exist. If so, what is the most likely transmitted codeword?

Solutions: The parity-check matrix of the code is

$$H = \begin{bmatrix} 1 & 1 & 0 & 1 & 0 & 0 \\ 0 & 1 & 1 & 0 & 1 & 0 \\ 1 & 0 & 1 & 0 & 0 & 1 \end{bmatrix}. \tag{8.13}$$

Multiplying the received sequence by H, we have

$$\mathbf{s} = H \cdot \begin{bmatrix} 1 \\ 1 \\ 1 \\ 0 \\ 1 \\ 1 \end{bmatrix} = \begin{bmatrix} 0 \\ 1 \\ 1 \end{bmatrix} \tag{8.14}$$

The syndrome is not all zero. Therefore, transmission error exists. Also, noting that $\mathbf{s}$ is identical to the third column of H, the third bit of the received sequence will be in error if there is only one error. The transmitted codeword is determined as 110011.

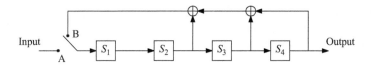

Figure 8.2   Sample implementation of cyclic encoder

Cyclic codes are another class of linear block codes. The codewords of cyclic codes are circular shifts of one another. Such special code structure facilitates the simple encoder and decoder implementation with linear-feedback shift registers (LFSR). A sample (n, k) cyclic encoder is shown in Figure 8.2. During the generation of each codeword, the switch will be first at position A, so that the information bits are shifted into the shift register. After k clock periods when all information bits are in the register, the switch is thrown to position B and the register starts generating output codeword. Note that the first k output bits will be the information bits and the remaining $n - k$ bits are parity-check bits calculated by the feedback logic. The decoder can be implemented with similar shift register.

Properly designed linear block codes can also achieve great performance. In fact, LDPC codes are a special type of linear block codes that can achieve performance close to the Shannon limit. LDPC codes were originally invented by Gallager in the 1960s and then reannounced in the 1990s after the appearance of turbo codes in 1993. LDPC uses parity-check matrix H with special structure. The fraction of nonzero entries of H is small, leading to the name of low density. Compared with conventional block codes, LDPC codes typically have much higher encoder and decoder complexity.

8.2.2   Performance benefit with coding

We now examine the performance benefit offered by channel coding. Let us consider a general (n, k) linear block code that can correct e errors in each codeword. We will study the codeword error probability of coded and uncoded systems, both of which use the same transmission power P_t to support the same information data rate R_b. It follows that the time to transmit each coded bit (channel bit) in coded system is R_c/R_b, whereas that for uncoded system is $1/R_b$. As such, the required bandwidth of coded system will be n/k times that of uncoded system. In addition, the amount of energy used to transmit each coded bit for the coded system will be $R_c P_t/R_b$, less than P_t/R_b for uncoded system. Therefore, the channel bits of coded system have higher probability to be detected in error. If P_b^u and P_b^c denote the channel bit error probability of uncoded system and coded system, respectively, then we have $P_b^u < P_b^c$.

Now let us consider the probability of codeword error for both systems. For uncoded system, each codeword consists of exactly k information bits. The codeword will be in error if at least one of its k bits is in error. Assuming channel bit errors are independent, the probability of codeword error for uncoded system is given by

$$P_w^u = 1 - (1 - P_b^u)^k. \tag{8.15}$$

The codeword of coded systems consists of n channel bits. Since the code can correct e errors, the codeword will be detected in error when at least $e + 1$ bits in the codeword are in error. Therefore, the probability of codeword error for coded system is calculated as

$$P_w^c = \sum_{i=e+1}^{n} \frac{n!}{i!(n-i)!}(1 - P_b^c)^{n-i}(P_b^c)^i. \tag{8.16}$$

Example: Performance improvement with coding

Let us consider the transmission of an information data stream with data rate 50 kbps over an AWGN channel with noise spectrum density $N_0 = 10^{-4}$ mW/Hz. The adopted modulation scheme is binary phase-shift keying (BPSK) with transmission power is 30 mW.

1. Determine the bit error probability of uncoded transmission.
2. Assume a coded system with a $(7, 4)$ single error-correction code is adopted. What is the bit error probability of the coded transmission?
3. Compare the probability of codeword error of coded and uncoded systems.

Solutions:

1. The bit error probability of BPSK over AWGN channel is calculated as $Q(\sqrt{2\gamma})$, where $\gamma = E_b/N_0$ is the received SNR per bit. The received SNR is determined as

$$\gamma = \frac{E_b}{N_0} = \frac{P_t/R_b}{N_0} = \frac{30 \times 10^{-3}}{10^{-7} \times 50 \times 10^3} = 6 = 7.8 \text{ dB}. \tag{8.17}$$

   As such, the bit error probability of uncoded transmission is $Q(\sqrt{12}) = 2.7 \times 10^{-4}$.

2. With coded transmission, the received SNR per channel bit becomes

$$\gamma = \frac{R_c P_t/R_b}{N_0} = (4/7) * 6 = 3.43 = 5.4 \text{ dB}. \tag{8.18}$$

   As such, the bit error probability of coded transmission is 4.4×10^{-3}, considerably higher than that of uncoded transmission.

3. The probability of codeword error of uncoded transmission is calculated as

$$P_w^u = 1 - (1 - 2.7 \times 10^{-4})^4 = 1.1 \times 10^{-3}. \tag{8.19}$$

   The probability of codeword error of coded transmission is determined, while noting that the code can correct single error as

$$P_w^c = \sum_{i=2}^{7} \frac{7!}{i!(7-i)!}(1 - 4.4 \times 10^{-3})^{7-i}(4.4 \times 10^{-3})^i = 4.0 \times 10^{-4}. \tag{8.20}$$

   The coded system enjoys lower codeword error probability and as such better reliability.

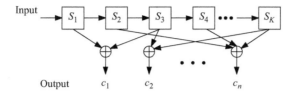

Figure 8.3   Structure of convolutional encoder

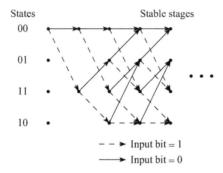

Figure 8.4   Trellis diagram of a (3, 1) convolution code with K = 3

8.2.3   *Convolutional codes and turbo codes*

Convolutional code is an example of nonblock codes. Different from block codes, the calculation of successive codewords with convolution codes usually shares common information bits. The convolutional encoder is typically implemented using a shift register with modulo-2 adders, as shown in Figure 8.3. k information bits are shifted into the register at a time. Then, an n-bit codeword is calculated for these k information bits. The resulting code rate R_c is equal to k/n. The length of shift register K, also known as the constraint length of convolutional code, is typically greater than k. As such, successive codewords are calculated using $K - k$ common information bits.

Convolutional codes are usually decoded with the Viterbi algorithm. First introduced by Viterbi in 1967, the Viterbi algorithm achieves the maximum likelihood decoding of convolutional codes, by exploring the trellis structure of the codewords. A sample trellis diagram is shown in Figure 8.4. In general, the diagram will have 2^{K-k} states and each state at stable stage will have 2^k incoming paths and 2^k outgoing paths. As such, the number of possible paths through the trellis grows exponentially with k, $K - k$, and the path length.

The Viterbi algorithm essentially looks for the most likely path corresponding to the received codewords. In particular, at each stage (after receiving each codeword), the decoder will compare the likelihood metrics of all paths entering a state and keep the one with the highest likelihood, usually referred to as the *surviving path*. The likelihood is determined by the similarity between the received codeword sequence

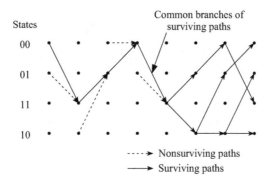

Figure 8.5   Maximum likelihood decoding of convolutional codes based on trellis diagram

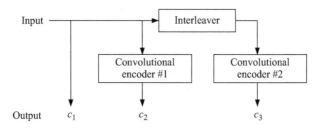

Figure 8.6   Structure of (3, 1) turbo encoder

and the codeword sequence of each path. The decoder will keep track of 2^{K-k} paths and their corresponding metrics. At each stage, the decoder will compute 2^k incoming path metrics to determine the surviving path for each state.

The Viterbi decoder will output the information bits when at a particular stage, all surviving paths traced back to a common state at an earlier stage. Specifically, the corresponding information bits can be determined using the common branches of the surviving paths. The concept of surviving path and common branch is illustrated in Figure 8.5. As there is no guarantee that the surviving paths will converge after a certain number of stages, the Viterbi decoder will introduce random decoding delay. One way to avoid such random decoding delay is to always decide the most likely path at n stages back. Another option is to periodically terminate the trellis with dummy information bits.

Turbo codes demonstrate to be a major breakthrough in the search for capacity approaching coding schemes for AWGN channels. Revealed in 1993 by Berrou, Glavieus, and Thitimajshima, turbo codes can come within a fraction of a decibel from the Shannon capacity limit. A typical turbo encoder consists of two parallel convolutional encoders separated by an interleaver, as shown in Figure 8.6. Note that the first output bit is the information bit. Therefore, the code is systematic. The convolutional encoders used here are typically recursive with a feedback path. Together with the interleaver, whose operation will be explained in latersection, the resulting design enables effective soft iterative decoding without excessive complexity.

8.3 Capacity of flat fading channels

Now let us consider a flat fading channel with input/output relationship over the ith symbol period given by

$$r_i = zs_i + n_i, \tag{8.21}$$

where z is the instantaneous complex channel gain. The instantaneous received SNR is given by

$$\gamma = \frac{gP_t}{N}, \tag{8.22}$$

where we use g to denote the instantaneous value of channel power gain $|z|^2$ for notation conciseness. Note that g is in general randomly varying over time, and as such, modeled a random variable with distribution functions depending on propagation scenarios. It follows that the instantaneous channel capacity of flat fading channel is given by

$$C_{\text{inst}} = B_s \log_2\left(1 + \frac{gP_t}{N}\right). \tag{8.23}$$

Since C_{inst} will vary over time, it cannot effectively characterize the rate upper limit of fading channels. Alternatively, ergodic (Shannon) capacity and capacity with outage are the commonly used capacity definition for fading channels.

8.3.1 Ergodic capacity

At any time instant, a flat fading channel can be viewed as an AWGN channel with capacity C_{inst} given in (8.23). Ergodic (Shannon) capacity of fading channel is defined as the statistical average of the instantaneous channel capacity over the distribution of fading channel power gain. Mathematically, we have

$$\overline{C} = \int_0^\infty B_s \log_2\left(1 + \frac{gP_t}{N}\right) p_{|z|^2}(g) dg, \tag{8.24}$$

where $p_{|z|^2}(\cdot)$ denotes the probability density function (PDF) of the channel power gain $|z|^2$.

Consider, for example, a Rayleigh fading wireless channel, where the PDF of the channel power gain $|z|^2$ is given by

$$p_{|z|^2}(g) = \frac{1}{2\sigma^2} \exp\left(-\frac{g}{2\sigma^2}\right), \tag{8.25}$$

where $2\sigma^2$ denotes the average channel power gain. Applying the definition of ergodic capacity, we have

$$\overline{C} = \int_0^\infty B_s \log_2\left(1 + \frac{gP_t}{N}\right) \frac{1}{2\sigma^2} \exp\left(-\frac{g}{2\sigma^2}\right) dg. \tag{8.26}$$

It can be shown, after carrying out integration, that the ergodic capacity of Rayleigh fading channel is given by

$$\overline{C} = B_s \log_2 e \exp\left(\frac{N}{2\sigma^2 P_t}\right) E_1\left(\frac{N}{2\sigma^2 P_t}\right), \tag{8.27}$$

where $E_1(\cdot)$ denotes the exponential integral function, defined by

$$E_1(x) = \int_1^\infty \frac{1}{t} \exp(-xt)dt. \tag{8.28}$$

According to the Jensen's inequality, we can show that

$$\int_0^\infty B_s \log_2\left(1 + \frac{gP_t}{N}\right) p_{|z|^2}(g)dg \leq B_s \log_2\left(1 + \frac{2\sigma^2 P_t}{N}\right), \tag{8.29}$$

where $\overline{g}$ denotes the average channel power gain, depending upon the path loss and shadowing effects. We can see that ergodic capacity of fading channel is always less than the Shannon capacity of the AWGN channel with the same average received SNR.

Example: Ergodic capacity of Rayleigh fading channels

Let us consider the digital transmission over a wireless channel that introduces Rayleigh fading. The transmit power is 10 mW, the channel bandwidth 30 kHz, the noise spectral density 10^{-9} W/Hz, and the average power gain considering path loss and shadowing effects -10 dB. Determine the ergodic capacity of the channel and compare it with that of a AWGN channel with the received SNR equal to the average received SNR of the fading channel.

Solutions: With the given information, we can determine the noise power at the receiver as 3×10^{-5} W $= 0.03$ mW. The average received SNR at the receiver can then be determined as

$$\overline{\gamma} = \frac{10 \times 0.1}{0.03} - 33.3. \tag{8.30}$$

We can calculate the ergodic capacity of the Rayleigh fading channel as $\overline{C} = 132$ kbps. The Shannon capacity of AWGN channel with received SNR of 33.3 is calculated as $B_s \log_2(1 + 33.3) = 153$ kbps, which is much greater than that of the fading channel.

Ergodic capacity is a widely used performance metric to characterize the transmission rate limit of fading channels. Meanwhile, achieving ergodic capacity is extremely difficult in real-world systems, even with the capacity achieving transmission schemes. Specifically, when the transmitter knows the instantaneous value of channel power gain $|z|^2$, the transmitter can adopt the corresponding capacity achieving transmission schemes to transmit at the rate that the channel can support. To implement such adaptation will however require that the transmitter and the receiver can adaptively switch among an infinite number of transmission schemes based on the channel condition. As $|z|^2$ is randomly varying over time, the transmitter needs to continuously adapt its transmission rate, which results in high transceiver complexity. In practical systems, such ideal continuous-rate adaptation is approximated with discrete rate adaptation as discussed in the following chapter.

Achieving ergodic capacity becomes even more challenging if the transmitter does not know the instantaneous value of $|z|^2$. In this case, the transmitter cannot adapt its transmission rate with the channel condition but transmit at a constant rate. Constant-rate transmission can achieve the ergodic capacity only if the codeword experience all possible fading channel condition, which implies extremely long code, and as such, excessive delay.

8.3.2 Capacity with outage

Capacity with outage is another commonly used capacity definition for fading channels, especially when the transmitter has no channel state information (CSI). The capacity with outage of a fading channel is defined as the maximum constant transmission rate that the channel can support for a given probability of outage, ε. Here, outage refers to the scenario that the channel condition is too poor to support the transmission rate with arbitrarily small error probability. Specifically, if the transmission rate is $C = B \log_2(1 + \gamma_0)$, where B is the channel bandwidth and γ_0 is a particular SNR value, outage occurs when the instantaneous received SNR γ is smaller than γ_0. Then, the outage probability is given by

$$P_{\text{out}}(\gamma_0) = \Pr[\gamma < \gamma_0]. \tag{8.31}$$

As such, for a given target outage probability, ε, the capacity with outage is calculated as

$$C_{\text{out}} = B \log_2(1 + \gamma_{\min}), \tag{8.32}$$

where

$$\gamma_{\min} = P_{\text{out}}^{-1}(\varepsilon). \tag{8.33}$$

Apparently, larger outage probability ε leads to larger value of $\gamma_{\min}$ and, as such, larger capacity with outage. On the other hand, the effective rate, i.e., average rate of correctly received data, given by

$$R_{\text{eff}} = (1 - P_{\text{out}})C_{\text{out}} = \Pr[\gamma > \gamma_{\min}]B \log_2(1 + \gamma_{\min}), \tag{8.34}$$

will reduce. In fact, there typically exists an optimal value of $\varepsilon/\gamma_{\min}$ that maximizes R_{eff}.

Consider a Rayleigh fading wireless channel, where the PDF of received SNR γ given by

$$p_\gamma(\gamma) = \frac{1}{\bar{\gamma}} \exp\left(-\frac{\gamma}{\bar{\gamma}}\right), \quad \gamma \geq 0, \tag{8.35}$$

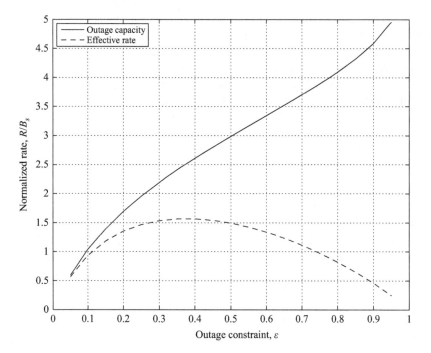

Figure 8.7   Effective rate of Rayleigh fading channel with constant rate transmission ($\bar{\gamma} = 10\,dB$)

where $\bar{\gamma} = 2\sigma^2 E_b/N_0$ denotes the average received SNR. The outage probability of the channel is calculated as

$$P_{\text{out}} = 1 - \exp\left(-\frac{\gamma}{\bar{\gamma}}\right). \tag{8.36}$$

To satisfy the target outage requirement of ε, γ_{min} can be determined as $\gamma_{\text{min}} = -\bar{\gamma}\ln(1 - \varepsilon)$. The capacity with outage is then calculated as

$$C_{\text{out}} = B\log_2(1 - \bar{\gamma}\ln(1 - \varepsilon)), \tag{8.37}$$

whereas the effective rate is calculated as

$$R_{\text{eff}} = (1 - \varepsilon)B\log_2(1 - \bar{\gamma}\ln(1 - \varepsilon)). \tag{8.38}$$

Figure 8.7 plots the capacity with outage and the effective rate, normalized by channel bandwidth, as function of ε. While the outage capacity is increasing monotonically with ε, the effective rate is a concave function of ε, indicating that an optimal value for ε exists.

8.3.3   Optimal power and rate adaptation

When the channel state information is available at the transmitter, the system can achieve ergodic channel capacity with continuous rate adaptation. Meanwhile, the channel capacity can be further enhanced with the application of power adaptation, leading to the so-called optimal power and rate adaptation (OPRA) capacity. The basic idea is to optimally set the transmission power level for different channel gain realization and then use capacity-achieving transmission scheme for the resulting instantaneous rate. More specifically, the optimal power adaptation strategy $P_t(g)$ should maximize

$$C = \int_0^\infty B \log_2 \left(1 + \frac{P_t(g)g}{N_0 B}\right) p_g(g) dg, \tag{8.39}$$

subject to the average transmit power constraint

$$\int_0^\infty P_t(g) p_g(g) dg \le \overline{P_t}. \tag{8.40}$$

The above optimization problem can be solved using the Lagrangian multiplier method (See Appendix A.4) as follows. The Lagrangian is formulated as

$$J = \int_0^\infty B \log_2 \left(1 + \frac{P_t(g)g}{N_0 B}\right) p_g(g) dg - \lambda \left(\int_0^\infty P_t(g) p_g(g) dg - \overline{P_t}\right). \tag{8.41}$$

Setting the partial derivative of J with respect to $P_t(g)$ to zero, we have

$$\frac{dJ}{dP_t(g)} = \left[\frac{B/\ln 2}{1 + gP_t(g)/N_0 B} \frac{g}{N_0 B} - \lambda\right] p_g(g) = 0. \tag{8.42}$$

Solving for $P_t(g)$ while noting the nonnegative transmit power constraint, we arrive at the optimal power adaptation policy as

$$P_t(g) = \begin{cases} N_0 B(1/g_T - 1/g), & g \ge g_T; \\ 0, & g < g_T, \end{cases} \tag{8.43}$$

where g_T is the cutoff value for the channel gain. After applying optimal power adaptation, the average power constraint becomes

$$\int_{g_T}^\infty N_0 B(1/g_T - 1/g) p_g(g) dg = \overline{P_t}, \tag{8.44}$$

from which g_T can be calculated numerically. The channel capacity with optimal power adaptation becomes

$$C^* = \int_{g_T}^\infty B \log_2(g/g_T) p_g(g) dg. \tag{8.45}$$

The resulting optimal power adaptation policy is the well-known water-filling solution. The behavior of the solution is illustrated in Figure 8.8. Specifically, the optimal policy suggests not to transmit if the channel power gain is too small, i.e., smaller than g_T. When the channel gain is larger, the transmit power is allocated proportional to the gap between horizontal line $1/g_T$ and curve $F(g) = 1/g$. Essentially,

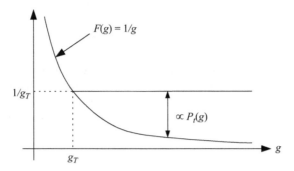

Figure 8.8   The water-filling solution of optimal power allocation

the larger the instantaneous channel power gain, the more the allocated transmission power. The basic principle is to use higher transmission power and transmit at a higher rate when the channel condition is better and to stop transmission when the channel condition is poor.

Example: Capacity with optimal power and rate adaptation

Consider the digital transmission over a special flat fading channel. The average transmission power $\overline{P_t}$ is 10 mW, the channel bandwidth 30 kHz, and the noise spectrum density is 10^{-8} W/Hz. The channel power gain g can take three possible values: $g_1 = 0.001$, $g_2 = 0.01$, and $g_3 = 0.05$ with probability 0.1, 0.6, and 0.3, respectively. Essentially, the channel gain follow a discrete distribution. Determine

1. the ergodic capacity (capacity with rate adaptation only) and
2. the OPRA capacity of the channel.

Solutions:

1. The ergodic capacity for such special channel can be calculated as

$$
C = \sum_i B \log_2 \left(1 + \frac{\overline{P_t} g_i}{N_0 B} \right) p_i, \tag{8.46}
$$

where p_i denotes the probability that the channel power gain g is equal to g_i, $i = 1, 2, 3$. After carrying out proper substitution and calculation, we have

$$
C = 30 \cdot 10^3 \cdot \left(\log_2 \left(1 + \frac{10^{-2} \cdot 0.001}{3 \cdot 10^{-4}} \right) \cdot 0.1 + \log_2 \left(1 + \frac{10^{-2} \cdot 0.01}{3 \cdot 10^{-4}} \right) \cdot 0.6 \right.
$$
$$
\left. + \log_2 \left(1 + \frac{10^{-2} \cdot 0.05}{3 \cdot 10^{-4}} \right) \cdot 0.3 \right) = 20.35 \text{ kbps} \tag{8.47}
$$

2. When the optimal power and rate adaptation is also applied, the channel capacity should be calculated using (8.50), with g_T determined from the average

power constraint. For the special wireless channel under consideration, the average transmit power constraint becomes

$$\sum_{g_i > g_T} N_0 B(1/g_T - 1/g_i)p_i = \overline{P_t}. \tag{8.48}$$

To determine g_T, we first assume $g_T < g_1 = 0.001$ and solve for g_T from the above equation. The solution from the average power constraint is $g_T = 0.005 > g_1$. The conflict implies that our initial assumption of $g_T < g_1$ is incorrect. We then assume $g_1 < g_T < g_2$, the average power constraint now becomes

$$N_0 B(1/g_T - 1/g_2)p_2 + N_0 B(1/g_T - 1/g_3)p_3 = \overline{P_t} \tag{8.49}$$

which leads to the correct $g_T = 0.0091$. No conflict occurs, which implies the g_T value is correct. Finally, the OPRA capacity is calculated as

$$C^* = B \log_2(g_2/g_T)p_2 + B \log_2(g_3/g_T)p_3 = 24.57 \text{ kbps}. \tag{8.50}$$

Clearly, the OPRA capacity is larger than the capacity achieved with rate adaptation only.

8.4 Capacity of selective fading channels

Now let us consider frequency-selective fading channels. The instantaneous frequency response of the channel over signal bandwidth is denoted by $H(f)$, which specifies the channel gain at frequency f over an incremental bandwidth of df. Therefore, the Shannon capacity of the channel can be shown to be given by

$$C = \int_{B_s} \log_2 \left(1 + \frac{|H(f)|^2 P(f)}{N_0}\right) df, \tag{8.51}$$

where $P(f)$ is the transmit power allocated to frequency f. Without channel state information at the transmitter, the transmitter will use equal power allocation on all frequency, i.e., $P(f) = P_t/B_s$, where P_t is the total transmission power. The Shannon capacity becomes

$$C_{\text{no CSIT}} = \int_{B_s} \log_2 \left(1 + \frac{|H(f)|^2 P_t}{B_s N_0}\right) df. \tag{8.52}$$

If the transmitter knows $H(f)$, it can apply transmit power allocation in the frequency domain to further improve capacity. The capacity of such selective fading channel with optimal power allocation can be determined by solving the following optimization problem

$$\max_{P(f)} \int_{B_s} \log_2\left(1 + \frac{|H(f)|^2 P(f)}{N_0}\right) df$$

s.t. $\int_{B_s} P(f)df \le P_t.$ (8.53)

Applying the Lagrangian multiplier method, we again arrive at a water-filling solution of the form

$$P(f) = \begin{cases} N_0(1/h_T - 1/|H(f)|^2), & |H(f)|^2 \ge h_T; \\ 0, & |H(f)|^2 < h_T, \end{cases}$$ (8.54)

where the threshold h_T satisfies the total transmit power constraint

$$\int_{B_s,|H(f)|^2 \ge h_T} P(f)df \le P_t.$$ (8.55)

The optimal power allocation solution over selective fading channel is illustrated in Figure 8.9. The resulting capacity is given by

$$C_{\text{CSIT}} = \int_{B_s,|H(f)|^2 \ge h_T} \log_2(|H(f)|^2/h_T)df.$$ (8.56)

In some scenarios, frequency-selective channel can be approximated by N parallel narrowband flat fading subchannels, each with channel gain $H_j, j = 1, 2, \ldots, N$. The bandwidth of each subchannel is approximately equal to B_c. As such, the number of parallel subchannels N is equal to B_s/B_c. Then, the Shannon capacity of the channel simplifies to

$$C = \sum_{j=1}^{N} B_c \log_2\left(1 + \frac{|H_j|^2 P_j}{N_0 B_c}\right).$$ (8.57)

The optimal power allocation policy becomes

$$P_j = \begin{cases} N_0(1/h_T - 1/|H_j|^2), & |H_j|^2 \ge h_T; \\ 0, & |H_j|^2 < h_T, \end{cases}$$ (8.58)

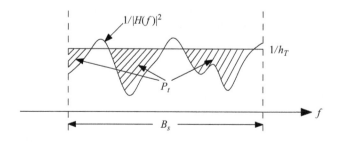

Figure 8.9   Optimal power allocation over frequency domain

where h_T satisfies the transmit power constraint

$$\sum_{j,|H_j|^2 \geq h_T} P_j \leq P_t. \tag{8.59}$$

leading to the channel capacity for the CSI at the transmitter scenario, given by

$$C_{\text{CSIT}} = \sum_{j,|H_j|^2 \geq h_T} \log_2\left(|H_j|^2/h_T\right). \tag{8.60}$$

The capacity of time-varying frequency-selective fading channels is generally unknown except for some special cases. The difficulty in the capacity analysis of such channels originates from the effect of intersymbol interference. If the channel can be approximated by parallel independent subchannels, then the capacity can be determined by first considering the individual subchannel capacity and then applying power allocation to them. The solution will involve a two-dimensional water filling.

8.5 Interleaving for fading channels

The channel coding schemes presented in previous section can readily apply to digital wireless transmission systems. Meanwhile, fading wireless channel tends to introduce long bursts of errors. Specifically, when the channel experiences deep fading, a long burst of errors will occur. Most of the coding schemes were designed for AWGN channel and typically cannot effectively correct long burst of errors. Codes that can correct error burst, however, are rather complex. Interleaving is an effective low-complexity solution to improve the performance of conventional coding schemes over fading wireless channel. The basic idea of interleaving is to spread the error burst over multiple codewords, such that each codeword has few errors, which can be corrected by the coding schemes presented in previous sections.

The structure of digital wireless transmission system with interleaving is shown in Figure 8.10. The codewords from the channel encoder are first processed by the interleaver, before being digitally modulated on to the carrier. The interleaver is

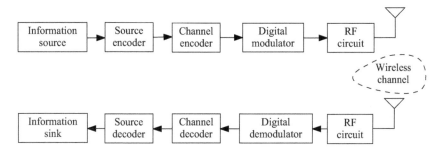

Figure 8.10 Digital wireless transmission system with interleaver

essentially an array of memory units. The bits of codewords will be written into the array by rows and read out from the array by columns. Meanwhile, the demodulator output is processed by the deinterleaver, which is an array of memory units, identical to that of the interleaver. The difference is that the received bits are written into the array by columns and read out by rows.

To further illustrate the interleaving operation, let us consider the digital transmission with an (n, k) block code. The sequence of coded bits are denoted by

$$\ldots, c_{11}, c_{12}, \ldots, c_{1n}, c_{21}, c_{22}, \ldots, c_{2n}, \ldots, c_{i1}, c_{i2}, \ldots, c_{in}, \ldots \qquad (8.61)$$

with $c_{i1}, c_{i2}, \ldots, c_{in}$ being the ith codeword. If these coded bits are directly transmitted over fading channel, some codewords may experience many errors, as the result of deep fade and will be decoded in error with high probability. With interleaving, the coded bits will be written into an interleaver array by rows and then read out by columns. Let us assume that the array has m rows and n columns, where $m > n$. The entries of the array after the coded bits are written in will be of the form

$$
\begin{array}{|c|c|c|c|}
\hline
c_{11} & c_{12} & \cdots & c_{1n} \\
\hline
c_{21} & c_{22} & \cdots & c_{2n} \\
\hline
\vdots & \vdots & \ddots & \vdots \\
\hline
c_{m1} & c_{m2} & \cdots & c_{mn} \\
\hline
\end{array}
\qquad (8.62)
$$

The output bit sequence read from the interleaver array becomes

$$c_{11}, c_{21}, \ldots, c_{m1}, c_{12}, c_{22}, \ldots, c_{m2}, \ldots, c_{1n}, c_{22}, \ldots, c_{mn} \qquad (8.63)$$

and will be transmitted over the channel.

The corresponding received bit sequence at the demodulator output is denoted by

$$c_{11}, c_{21}, \ldots, c_{m1}, c'_{12}, c'_{22}, \ldots, c'_{n2}, c_{(n+1)2}, \ldots c_{m2}, \ldots, c_{1n}, c_{22}, \ldots, c_{mn} \qquad (8.64)$$

Here, we assume that the fading channel causes n consecutive coded bit errors, which affect the detection of $c_{12}, c_{22}, \ldots, c_{n2}$ and lead to received bits $c'_{12}, c'_{22}, \ldots, c'_{n2}$. Without interleaving, these errors will affect one or two codewords, rendering their correct decoding very difficult. With interleaving, each of these n consecutive bit errors will affect n different codewords. In particular, the deinterleaver array after writing in the received bits becomes

$$
\begin{array}{|c|c|c|c|c|}
\hline
c_{11} & c'_{12} & c_{13} & \cdots & c_{1n} \\
\hline
c_{21} & c'_{22} & c_{23} & \cdots & c_{2n} \\
\hline
\vdots & \vdots & \vdots & \ddots & \vdots \\
\hline
c_{n1} & c'_{n2} & c_{n3} & \cdots & c_{nn} \\
\hline
c_{(n+1)1} & c_{(n+1)2} & c_{(n+1)3} & \cdots & c_{(n+1)n} \\
\hline
\vdots & \vdots & \vdots & \ddots & \vdots \\
\hline
c_{m1} & c_{m2} & c_{m3} & \cdots & c_{mn} \\
\hline
\end{array}
\qquad (8.65)
$$

and the output sequence from the deinterleaver is

$$c_{11}, c'_{12}, c_{13}, \ldots, c_{1n}, c_{21}, c'_{22}, c_{23}, \ldots, c_{2n}, \ldots, c_{n1}, c'_{n2}, \ldots, c_{nn}, \ldots, c_{mn} \quad (8.66)$$

Each of the first n codewords contains one bit errors from the error burst. If the adopted block code can correct a single bit error, then these codewords can be detected correctly.

Similarly, we can show that if the error burst spans more than m bits but less than $2m$ bits, then some codewords will have two bit errors, which can be corrected with double-error-correcting codes. Therefore, the number of rows of the interleaver/deinterleaver array, m, also known as the *interleaving depth*, is an important design parameter. Specifically, if the maximum length of the burst error is K, then m should be chosen to be greater or equal to K, to avoid double errors from the same error burst in a codeword. The maximum length of error burst due to fading typically depends on the channel coherence time of the channel. On the other hand, larger m would necessarily increase the decoding delay of the codewords. With interleaving, the first codeword can be decoded only after $(m-1)n+1$ bits of the interleaver content have been received, which linearly increases with m. As such, there is a tradeoff between the requirement on error correction capability and decoding delay associated in the design of interleaver depth m.

Example: Interleaver design over flat fading

Consider the digital transmission over a frequency flat fading channel. The channel coherence time is 5 ms. The coded information bits are transmitted at a rate of 25 kbps. The system adopts a linear binary block code with codeword length 7 that is capable of correcting single errors. Determine the minimum interleaver depth and the maximum decoding delay if we want each codeword to contain at most one bit error from the same error burst. What if the code can correct double errors?

Solutions: The duration of deep fade for these channel is approximately $T_c = 5$ ms. As such, the maximum length of the error burst can be estimated as $T_c \cdot R_b = 5 \times 25 = 125$. To ensure each codeword contains at most one bit error from a single error burst, the interleaver depth m should be at least 125. The interleaver array will be of the dimension of 125 by 7. The decoding delay of the first codeword written into the array is

$$\frac{(m-1)n+1}{R_s} = \frac{124 \times 7 + 1}{25} = 34.76 \text{ ms.} \quad (8.67)$$

If the code adopted can correct double error, then we can allow each codeword contain two bit errors. The interleaver depth m should be at least $\lceil 125/2 \rceil = 63$. The decoding delay of the first codeword becomes

$$\frac{62 \times 7 + 1}{25} = 17.4 \text{ ms.} \quad (8.68)$$

Coding with interleaving also applies to digital multicarrier transmission over frequency-selective fading channels. The basic idea is to transmit coded bits over different subchannels such that only few bits in each codeword suffer bad subchannel. If most coded bits are received correctly, then the errors caused by bad subchannels can be corrected. When the channel coherence bandwidth, B_c, is large, the fading experienced by neighboring subchannels will be highly correlated. In such scenario, the interleaving operation becomes again essential to maintain the benefit of channel coding. The interleaving depth should be designed based on the channel coherence bandwidth as well as the constellation size of the modulation scheme, M. With OFDM implementation, the modulated symbols in each block are transmitted on consecutive subcarriers. As such, if each modulated symbol carries $n = \log_2 M$ coded bits and there are l subcarriers in each B_c, then the interleaver depth should be at least $n \cdot l$ for the coded bits of a codeword to experience independent subchannel realizations.

Example: Interleaver design over selective fading

Consider the digital transmission over a selective fading channel. The channel bandwidth is 2 MHz and the coherence bandwidth is 150 kHz. The transmission system adopts 16-QAM modulation scheme and an OFDM implementation with block size 64. Determine the minimum interleaver depth if we want each bit of a codeword experiences different fading. What if block size is increased to 128?

Solutions: The subchannel bandwidth with the adopted OFDM implementation is $B_s/N = 2,000/64 = 31.25$ kHz. The number of subchannels over one channel coherence time is $150/31.25 = 4.8 \approx 5$. Since each modulated symbol with 16-QAM carries 4 bits of information, the minimum interleaver depth will be 20. If the block size increases to 128, then the number of subchannels over one channel coherence time becomes $150/15.625 = 9.6 \approx 10$. The interleaver depth becomes 40.

Coding with interleaving can also achieve a certain diversity benefit. Specifically, with interleaving, the coded bits corresponding to the same codeword will be separated by $m - 1$ bits. If the transmission time of these $m - 1$ bits is longer than the channel coherence time T_c, then different coded bits of the same codeword will experience independent channel condition. Since these coded bits are related to same information bits, the diversity gain can be obtained when the codewords are used to recover the information bits. Similarly, if the bandwidth required to transmit these $m - 1$ bits is greater than the channel coherence bandwidth B_c, then the coded bits transmitted over selective fading channels will experience independent subchannel fading. Certain diversity gain will be extracted during the decoding process.

8.6   Further readings

The derivation of Shannon capacity can be found in [1]. Several textbooks are dedicated to information theory [2,3]. Further discussion about various block codes and convolutional codes and their performance can be found in [4, Chapter 8], [5, Chapter 8]. Reference [6, Chapter 4] provides further discussion on capacity definition for fading wireless channels.

Problems

1. Consider an AWGN channel with bandwidth 100 kHz. Determine the minimum required E_b/N_0 value such that the channel can support error-free transmission of information stream with data rate of 500 kbps.

2. Let us consider a (5, 2) block code that generates codewords for each bit pair in the following fashion.

$$00 \Longrightarrow 00000; \qquad (8.69)$$

$$01 \Longrightarrow 10101;$$

$$11 \Longrightarrow 11110;$$

$$10 \Longrightarrow 01011.$$

(i) Determine the minimum distance between codewords. (ii) If 11010 is received, what is the most likely transmitted information bit pair? (iii) Assume that code-word 10101 is transmitted, determine the error patterns **e** that can be corrected with maximum likelihood decoding principle.

3. A (6, 3) parity-check code has the following parity-check matrix

$$H = \begin{bmatrix} 1 & 1 & 0 & 1 & 0 & 0 \\ 1 & 1 & 1 & 0 & 1 & 0 \\ 0 & 1 & 1 & 0 & 0 & 1 \end{bmatrix}. \qquad (8.70)$$

Determine the generator matrix and all possible codewords.

4. Let us consider the transmission of an information data stream with data rate 200 kbps over an AWGN channel with noise spectrum density $N_0 = 10^{-4}$ mW/Hz. The adopted modulation scheme is BPSK with transmission power is 30 mW. Assume a coded system with a (6, 3) single error-correction code adopted. What is the bit error probability and codeword error probability of the coded transmission?

5. Consider the digital transmission over a special flat fading channel. The average transmission power $\overline{P_t}$ is 10 mW, the channel bandwidth 30 kHz, and the noise spectrum density is 10^{-8} W/Hz. The channel power gain g can take three possible values: $g_1 = 0.001$, $g_2 = 0.01$, and $g_3 = 0.05$ with probability 0.1, 0.4, and 0.5, respectively. Determine the outage capacity of the channel for the outage probability of (i) 0.2 and (ii) 0.4, and (iii) 0.6.

6. Let us consider the digital transmission over a Rayleigh fading channel. The transmit power is 10 mW, the channel bandwidth 50 kHz, the noise spectral density 10^{-8} W/Hz, and the average power gain considering path loss and shadowing effects -15 dB. Determine the ergodic capacity and OPRA capacity of the channel.

7. Consider a frequency-selective channel that can be converted into five subchannels of bandwidth 500 kHz each. The instantaneous power gain of the subchannels $|H_i|^2$ are 0.1, 0.25, 0.12, 0.17, and 0.06. The total transmit power is 10 mW and the noise power spectral density is 10^{-8} W/Hz. Determine the instantaneous capacity of the channel with optimal power allocation.

8. Consider the digital transmission over a frequency flat fading channel. The channel coherence time is 7 ms. The coded information bits are transmitted at a rate of 25 kbps. The system adopts a (11, 5) binary block code. Due to decoding delay constraint, the interleaver depth is at most 120. Determine the maximum information data rate if we want each codeword contain at most one bit error from the same error burst.

Bibliography

[1] C. E. Shannon and W. Weaver, *The Mathematical Theory of Communication*, Urbana, IL: University of Illinois Press, 1949.

[2] R. G. Gallager, *Information Theory and Reliable Communication*, New York, NY: Wiley, 1968.

[3] T. Cover and J. Thomas, *Elements of Information Theory*, New York, NY: Wiley, 1991.

[4] J. G. Proakis, *Digital Communications*, 4th ed. New York, NY: McGraw-Hill, 2001.

[5] G. L. Stüber, *Principles of Mobile Communications*, 2nd ed. Norwell, MA: Kluwer Academic Publishers, 2000.

[6] A. Goldsmith, *Wireless Communications*, New York, NY: Cambridge University Press, 2005.

Chapter 9

Channel adaptive transmission

Wireless channel introduces a time-varying gain to the transmitted signal. The dynamic range of channel power gain can be as much as 30 dB due to the fading effect. Adaptive transmission can achieve high spectral and power efficiency with guaranteed error rate performance over wireless fading channels. As such, adaptive transmission becomes an essential technology to meet the increasing demand for highly spectrum efficient wireless transmission. This chapter studies channel adaptive transmission techniques over frequency flat fading channels. We first present the basic idea of channel adaptive wireless transmission. We then separately discuss two classes of adaptive transmission technologies, namely rate adaptation and power adaptation. A joint discrete rate and continuous power adaptation design is also presented together with several implementation issues associated with channel adaptive transmission.

9.1 Adaptive transmission

The basic idea of adaptive transmission is to vary the transmission schemes/parameters, such as modulation scheme, coding scheme/rate, or transmitting power, with the prevailing fading channel conditions. The system will exploit more favorable channel conditions with higher rate transmission at lower power levels and response to channel degradation with reduced data rate or increased power level. As a result, the overall system throughput is maximized with controlled transmit power consumption while maintaining a certain desired reliability.

The generic structure of a digital wireless transmitter is shown in Figure 9.1. Information bits are first encoded by channel encoder, which adds controlled redundancy to protect the information bits. Different channel coding schemes will entail different trade-offs between reliability, characterized by the error rate of information bits, and efficiency, measured by the percentage of redundancy introduced during the coding process. Coding rate, defined as the average number of information bits carried in each channel bit, usually denoted by $R_c \leq 1$, serves as a quantitative measure for coding efficiency. Larger R_c implies better coding efficiency, but less protection to information bits. If $R_c = 1$, then no redundancy bits are added at all, which corresponds to the no-channel-coding scenario.

The coded bits at the output of the channel encoder will then be processed by the modulator. The modulator essentially maps coded bits to data symbols that are suitable

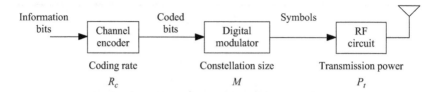

Figure 9.1   Generic structure of a digital wireless transmitter

for transmission over the wireless channel. Similar to channel encoding, different modulation schemes lead to different trade-off between reliability and transmission efficiency. Most digital wireless transmitters use sinusoidal waveforms of fixed duration, equal to symbol period T_s, to carry coded bits. The number of coded bits that a modulated symbol can carry depends on the number of available distinct symbols, also known the *constellation size*, of the modulation scheme. We denote the constellation size by M. An M-ary modulation scheme transmits $\log_2 M$ bits over each symbol period T_s. For the same symbol rate and transmission power, larger M leads to higher information data rate, but less reliable transmission, i.e., higher symbol error rate.

Finally, the modulated symbols are processed by the RF circuit before being transmitted from the antenna. Major RF operation includes carrier modulation and power amplification to a desired power level of P_t. The wireless transmitter may adapt the coding rate R_c, the constellation size M, and/or the transmission power P_t with the fading channel condition. Varying R_c/M leads to the so-called rate adaptation schemes, whereas power adaptation scheme changes P_t based on the channel realization. Certain implementation varies both transmission rate and transmission power. In the following, we will explain the basic design principles of these schemes by assuming a flat fading channel model. Specifically, the wireless channel introduces a complex channel gain z. The instantaneous received signal-to-noise ratio (SNR) is then related to z as

$$\gamma = \frac{P_t|z|^2}{N_0 B_s},\tag{9.1}$$

where N_0 is the power spectrum density of additive white Gaussian noise (AWGN) collected by the receiver, and B_s is the channel bandwidth.

9.2   Rate adaptation

The information data rate of digital wireless transmission, defined as the average number of information bits transmitted over each symbol period T_s, can be calculated as

$$R_b = R_c \cdot \log_2 M/T_s,\tag{9.2}$$

where $\log_2 M/T_s$ essentially characterizes the number of coded bits carried by each symbol. Note that the symbol period T_s is determined by the channel bandwidth available for transmission. Based on the Nyquist criterion, the minimum symbol period for

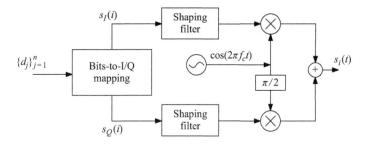

Figure 9.2 Structure of linear modulator

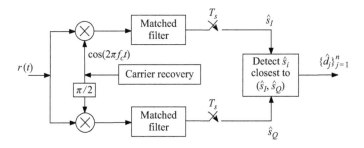

Figure 9.3 Structure of linear demodulator

the transmission over a channel with bandwidth B_s to avoid intersymbol interference is $1/B_s$. As the channel bandwidth is typically fixed during transmission, we assume in the following discussion that the symbol period T_s is a constant.

With rate adaptation schemes, the transmitter adapts the information data rate R_b, by varying the coding rate R_c and/or the constellation size M, with the instantaneous channel condition, while using constant transmit power P_t. Such rate adaptation schemes are often referred to as adaptive modulation and coding (AMC) in various wireless standards. The basic premise of such adaptation schemes is that the channel encoder and digital modulator should be reconfigurable on the fly. Since most channel encoders and decoders are implemented using digital circuit, coding rate adaptation is readily implementable. Adapting the constellation size typically involves the changing of modulation schemes. Fortunately, most linear modulation schemes, including phase shift keying (PSK) and quadrature amplitude modulation (QAM) schemes, share similar modulator and demodulator structure, as shown in Figures 9.2 and 9.3. Different linear modulators differ only in the coded bits to I/Q components mapping process, whereas the demodulators use different decision rules for different modulation schemes. As such, we can also adjust the constellation size with linear modulation schemes.

Typically, a finite set of coding schemes and/or modulation types are used in real-world systems. As such, the transmission rate is adapted in a discrete manner with a finite set of possible values. Table 9.1 shows a sample set of modulation and coding schemes (MCS) and their corresponding information data rate, in bits

Table 9.1   Sample modulation and coding schemes (MCS) for rate adaptation

Index	Modulation type	Coding scheme	Data rate (bits/T_s)
1	QPSK	1/2 convolution	1
2	QPSK	2/3 convolution	4/3
3	16-QAM	1/2 convolution	2
4	16-QAM	2/3 convolution	8/3
5	64-QAM	1/2 convolution	3
6	64-QAM	2/3 convolution	4

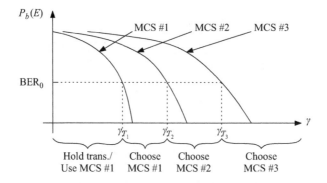

Figure 9.4   Modulation and coding scheme selection

per symbol period. Here, the convolution codes with rate 1/2 and 2/3 are used in combination with quadrature phase shift keying (QPSK), 16-QAM, and 64-QAM modulation schemes. The information data rate takes six different values, ranging from 1 to 6 bits/T_s. In general, the higher the information data rate, the higher the bit error rate (BER) for the given channel realization.

The MCS schemes are selected based on the instantaneous quality of the wireless channel, characterized by the received SNR γ. Typically, the MCS scheme that achieves the highest spectrum efficiency while maintaining acceptable instantaneous BER performance is chosen. This idea is further illustrated in Figure 9.4, where we plot the instantaneous BER of three different MCS schemes as the function of the received SNR γ. Here, we assume that MCS scheme 3 achieves the highest spectrum efficiency but with the worst BER performance, whereas MCS scheme 1 achieves the lowest spectrum efficiency but with the best BER performance. The target instantaneous BER value is denoted by BER_0. The required SNR value for each MCS scheme to satisfy the target BER value, also referred to as SNR threshold, is marked as $\gamma_{T_1} < \gamma_{T_2} < \gamma_{T_3}$. As such, MCS scheme 2 is chosen when received γ falls into the interval of $(\gamma_{T_2}, \gamma_{T_3})$. Note that MCS scheme 2 achieves higher spectrum efficiency even though using either MCS scheme 1 or MCS scheme 2 would satisfy the instantaneous BER requirement.

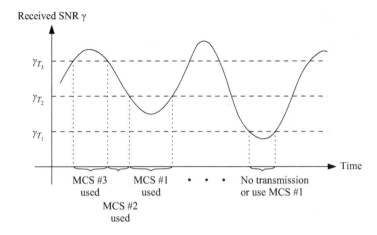

Figure 9.5 Sample operation of rate adaptive transmission

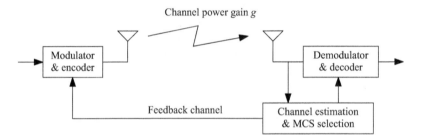

Figure 9.6 Structure of rate adaptive transmission system

It may happen in the worst case scenario that the instantaneous received SNR is less than γ_{T_1}. Essentially, none of the available MCS schemes can satisfy the target BER requirement. In this case, the system may either choose to hold the transmission until the channel condition improves or to transmit in violation of the target BER requirement. Figure 9.5 illustrates the sample operation of rate adaptation with three MCS schemes. The MCS scheme selection is typically carried out at the receiver who can readily estimate the received SNR. The receiver will compare the estimated receive SNR with the SNR threshold γ_{T_n}s for different MCS schemes. After determining the suitable MCS scheme based on the comparison results, the receiver feeds back the index of the selected MCS scheme to the transmitter. Then, both the transmitter and the receiver are reconfigured to communicate with the chosen MCS scheme. Such implementation strategy entails a feedback load of only $\lceil \log_2 N + 1 \rceil$ bits, where N is the total number of available MCS schemes. The system structure of rate adaptative transmission is shown in Figure 9.6.

9.2.1 Adaptive modulation

The SNR threshold γ_{T_n}s play a critical role in the operation of adaptive transmission systems. These thresholds are determined based on the instantaneous BER requirement. Specifically, γ_{T_n} is the smallest SNR value required by the nth MCS scheme to achieve an instantaneous BER value no greater than the target BER value, BER_0. Calculating the exact value of γ_{T_n}s requires the analytical expression of the instantaneous BER as function of the received SNR, which is very challenging to obtain for coded transmission system. To better explain the process of determining γ_{T_n} values, we now consider constant-power uncoded adaptive M-QAM system in this subsection. In particular, the system adaptively selects one of N different M-QAM modulation schemes based on the fading channel condition without applying channel coding.

Different M-QAM schemes differ by their constellation sizes. For squared M-QAM schemes, the constellation sizes are $M = 2^n$, $n = 1, 2, \ldots, N$, with size 2^n corresponding to an information data rate of n bits per symbol period. According to the Nyquist criterion, such information data rate translates to a spectral efficiency of approximately n bps/Hz. Again, the modulation schemes are chosen to achieve the highest spectral efficiency while maintaining the instantaneous error rate below the target BER_0 value. Specifically, the value range of the received SNR γ is divided into $N + 1$ regions, with threshold values denoted by $0 < \gamma_{T_1} < \gamma_{T_2} < \cdots < \gamma_{T_N} < \infty$. When the received SNR γ falls into the nth region, i.e., $\gamma_{T_n} \leq \gamma < \gamma_{T_{n+1}}$, the M-QAM scheme with constellation size 2^n will be selected for transmission.

Let $\text{BER}_n(\gamma)$ denote the instantaneous BER of squared M-QAM over AWGN channel with SNR γ. To satisfy $\text{BER}_n(\gamma) \geq \text{BER}_0$, we need

$$\gamma \geq \gamma_{T_n} = \text{BER}_n^{-1}(\text{BER}_0), \tag{9.3}$$

where $\text{BER}_n^{-1}(\cdot)$ is the inverse BER expression. It has been shown that instantaneous BER of M-QAM with SNR γ can be approximately calculated as

$$\text{BER}_n(\gamma) = \frac{1}{5} \exp\left(-\frac{3\gamma}{2(2^n - 1)}\right), \quad n = 1, 2, \ldots, N. \tag{9.4}$$

As such, the threshold values γ_{T_n} can be calculated as

$$\gamma_{T_n} = -\frac{2}{3} \ln(5\,\text{BER}_0)(2^n - 1); \quad n = 1, 2, \ldots, N. \tag{9.5}$$

Example: Adaptive modulation scheme design

Consider an adaptive modulation system using three modulation schemes, QPSK, 16-QAM, and 64-QAM. The modulation schemes are chosen adaptively based on the instantaneous channel condition to satisfy the instantaneous BER requirement of 10^{-4}. Determine the mode of operation of the adaptation modulation scheme.

Solutions: We need to determine the value range of received SNR for each modulation scheme. Applying the approximate BER expression for M-QAM given above, the threshold for QPSK, or equivalently 4-QAM, is determined as

$$\gamma_{T_1} = -\frac{2}{3}\ln(5 \cdot 10^{-4})(4 - 1) = 15.2 = 11.8 \text{ dB}. \tag{9.6}$$

Similarly, the thresholds for 16-QAM and 64-QAM can be determined as $\gamma_{T_2} = 76 = 18.8$ dB and $\gamma_{T_3} = 319.2 = 25$ dB, respectively. As such, the adaptive modulation scheme can be implemented as following: When the received SNR γ is greater than 25 dB, the system uses 64-QAM; When γ is between 18.8 and 25 dB, the system uses 16-QAM; When γ is between 11.8 and 18.8 dB, QPSK is used. If γ is less than 11.8 dB, the system may choose to hold the transmission and wait until the channel quality is better. Alternatively, the system may still transmit with QPSK, at the cost of not satisfying the instantaneous BER requirement.

9.2.2 Performance analysis over fading channels

The performance of rate adaptation system can be evaluated in terms of the average spectral efficiency and the average BER. In particular, the average spectral efficiency of the rate adaptation system under consideration can be calculated as

$$\eta = \sum_{n=1}^{N} R_n \cdot P_n, \tag{9.7}$$

where R_n is the spectral efficiency of the nth MCS scheme, in bps/Hz, and P_n is the probability of using the nth MCS scheme. Note that R_n also characterizes the information data rate of the nth MCS scheme. As such, η can also be interpreted as the average information data rate of rate adaptive transmission system. With the application of appropriate fading channel model, P_n can be calculated using the distribution function of the received SNR as

$$P_n = \int_{\gamma_{T_n}}^{\gamma_{T_{n+1}}} p_\gamma(x)dx, \tag{9.8}$$

where $p_\gamma(\cdot)$ denotes the probability density function (PDF) of the received SNR γ.

We now derive the average BER of rate adaptive transmission systems. Applying the total probability theorem, the average BER can be calculated as

$$\langle \text{BER} \rangle = \sum_{n=1}^{N} \text{Pr}[\text{bit transmitted using MCS } n] \cdot \overline{\text{BER}}|_n, \tag{9.9}$$

where Pr[bit transmitted using MCS n] denotes the probability that an information bit is transmitted using nth MCS scheme, and $\overline{\text{BER}}|_n$ denotes the average bit error probability given that MCS n is used to transmit the bit. Pr[bit transmitted using MCS n] can be calculated as the ratio of the number of bits transmitted using MCS scheme n

over the average number of bits transmitted per symbol period. Mathematically, we have

$$\Pr[\text{bit transmitted using MCS } n] = \frac{R_n P_n}{\sum_{i=1}^{N} R_i P_i} = \frac{R_n P_n}{\eta}, \tag{9.10}$$

where R_n is the number bits that are transmitted in each symbol period when MCS scheme n is selected. Given a bit is transmitted using MCS scheme n, the average BER can be calculated as

$$\overline{\text{BER}}|_n = \frac{\int_{\gamma T_n}^{\gamma T_{n+1}} \text{BER}_n(\gamma) p_\gamma(x) dx}{P_n}, \tag{9.11}$$

where $\text{BER}_n(\gamma)$ is the instantaneous BER of MCS scheme n with SNR γ. As such, the numerator is the joint probability that a bit is transmitted using MCS scheme n and the bit is detected in error. After proper substitution, the average BER of the rate adaptive transmission system can be calculated as

$$\langle \text{BER} \rangle = \sum_{n=1}^{N} \frac{R_n P_n}{\eta} \frac{\int_{\gamma T_n}^{\gamma T_{n+1}} \text{BER}_n(\gamma) p_\gamma(x) dx}{P_n} \tag{9.12}$$

$$= \frac{1}{\eta} \sum_{n=1}^{N} R_n \overline{\text{BER}}_n,$$

where $\overline{\text{BER}}_n$ denotes the joint probability that MCS scheme n is used, and there is a bit error, given by

$$\overline{\text{BER}}_n = \int_{\gamma T_n}^{\gamma T_{n+1}} \text{BER}_n(x) p_\gamma(x) dx. \tag{9.13}$$

The weight R_n is introduced to account for the fact that different modulation schemes transmit different number of bits in each symbol period. More bits will experience the average error rate of higher order modulation schemes.

Example: Performance analysis of adaptive modulation system

Let us consider the digital wireless transmission over a Rayleigh flat fading channel with average received SNR $\overline{\gamma} = 20$ dB. The adaptive modulation scheme (see previous example) uses QPSK, 16-QAM, and 64-QAM over the SNR region [11.8, 18.8) dB, [18.8, 25) dB, and [25, ∞) dB, respectively, and stops transmission when the received SNR is less than 11.8 dB. Compare the performance of such adaptive transmission system with a conventional transmission system that always uses the QPSK modulation scheme regardless of the channel realization, in terms of

1. spectral efficiency and
2. instantaneous error rate.

Solutions:

1. The spectral efficiency of the conventional system with QPSK is always 2 bits/s/Hz. The spectral efficiency of the adaptive modulation system is

$$\eta = 2\Pr[15.2 \le \gamma < 76] + 3\Pr[76 \le \gamma < 319.2] + 4\Pr[319.2 \le \gamma]$$

$$= 2.2267 \text{ bps/Hz}. \tag{9.14}$$

Clearly, although the adaptive transmission will not transmit when $\gamma < 11.8$ dB, it still achieves much higher average spectral efficiency than conventional QPSK system, by better exploring good channel condition.

2. The instantaneous error rate of the conventional QPSK system will vary with the instantaneous received SNR, as the instantaneous BER of QPSK is given by $Q(\sqrt{2\gamma})$. With probability $\Pr[\gamma < 15.2] = 0.141$, the instantaneous BER will be larger than 10^{-4}. On the other hand, the instantaneous BER of adaptive transmission system will be always below 10^{-4} during transmission.

9.3 Power adaptation

With power adaptation schemes, the transmit power P_t is adjusted with the instantaneous channel condition, while the information data rate R_b remains constant, depending on the adopted MCS. A typical objective of power adaptation is to mitigate the effect of channel power gain variation due to fading and to maintain constant received SNR for a certain BER target. While definitely suboptimal, such objective leads to low-complexity transceiver structure. Note that the received SNR is related to the transmission power as

$$\gamma = \frac{P_t \cdot g}{N}, \tag{9.15}$$

where $g = |z|^2$ denotes the channel power gain, and N is the noise power at the receiver. The power adaptation policy is typically specified by a function of g, denoted by $P_t(g)$.

9.3.1 Full channel inversion

The objective of full channel inversion power adaptation strategy is to maintain a constant received SNR, denoted by γ_t, over all possible channel realization. Specifically, the transmit power $P_t(g)$ is set such that

$$\frac{P_t(g) \cdot g}{N} = \gamma_t. \tag{9.16}$$

The transmit power is proportional to the inverse of the channel power gain, i.e.,

$$P_t(g) = \frac{\gamma_t \cdot N}{g}. \tag{9.17}$$

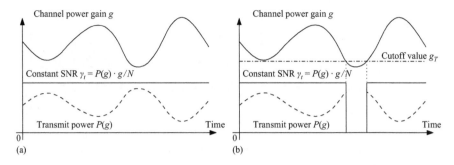

Figure 9.7   *Operation of power adaptive transmission: (a) full channel inversion and (b) truncated channel inversion*

Meanwhile, the power adaptation strategy should satisfy the average power constraint that entails

$$\int_0^{\infty} P_t(g)p_g(g)dg \leq \overline{P_t}, \tag{9.18}$$

where $\overline{P_t}$ denotes the maximum average transmit power. After proper substitution and some manipulation, we can show that γ_t is upper bounded as

$$\gamma_t \leq \frac{\overline{P_t}}{N \int_0^{\infty} (1/g)p_g(g)dg} = \frac{\overline{P_t}}{N \cdot \mathbf{E}[1/g]}. \tag{9.19}$$

where $\mathbf{E}[\cdot]$ denotes the statistical averaging operation.

Figure 9.7(a) illustrates the operation of full channel inversion power adaptation strategy. In particular, the channel power gain variation over time is plotted together with the corresponding instantaneous transmission power, which is inverse proportional to each other. The constant received SNR is indicated by a horizontal line. The disadvantage of full channel inversion is that the system uses very high transmit power when the channel gain g takes small values. If the probability of g being small is relatively high, then constant target SNR γ_t that the system can maintain will be small. In fact, for Rayleigh fading environment, where g follows exponential distribution, $\mathbf{E}[1/g]$ can be shown to be equal to infinity. Full channel inversion cannot maintain positive constant SNR for Rayleigh fading channel. We can also observe that the operation strategy of full channel inversion is kind of opposite to the optimal water-filling solution for capacity optimization over fading channels. As such, we can expect intuitively that better performance can be achieved with improved power adaptation strategies.

Example: Full channel inversion

Consider the power adaptive transmission over a special flat fading channel. The channel power gain g can take three possible values: $g_1 = 0.001$, $g_2 = 0.01$, and $g_3 = 0.05$ with probability 0.1, 0.6, and 0.3, respectively. Essentially, the channel gain follows a discrete distribution. The average transmission power $\overline{P_t}$ is 10 mW, the channel bandwidth 30 kHz, and the noise spectrum density is 10^{-8} W/Hz. Determine maximum constant received SNR that the system can maintain at the receiver with full channel inversion power adaptation strategy and the corresponding power adaptation scheme.

Solutions: We first determine $\mathbf{E}[1/g]$ as

$$\mathbf{E}[1/g] = 0.1/0.001 + 0.6/0.01 + 0.3/0.05 = 166. \tag{9.20}$$

The maximum target SNR under the average transmit power constraint is

$$\gamma_t = \frac{10^{-2}}{30 \times 10^3 \times 10^{-8} \times 166} = 0.2, \tag{9.21}$$

which is quite small. Accordingly, the power adaptation policy to maintain such maximum constant SNR is

$$P_t(g) = \begin{cases} \frac{\gamma_t B_s N_0}{g_1}, & g = g_1; \\ \frac{\gamma_t B_s N_0}{g_2}, & g = g_2; \\ \frac{\gamma_t B_s N_0}{g_3}, & g = g_3. \end{cases} \tag{9.22}$$

$$= \begin{cases} 60, & g = 0.001; \\ 6, & g = 0.01; \\ 1.2, & g = 0.05. \end{cases}$$

9.3.2 Truncated channel inversion

With the truncated channel inversion power adaptation strategy, the transmission system maintains a constant received SNR γ_t when the channel gain g is large enough. When the channel gain is too small, the transmitter will hold the transmission until the channel condition improves. The transmit power $P_t(g)$ with truncated channel inversion is set as

$$P_t(g) = \begin{cases} \frac{\gamma_t \cdot N}{g}, & g \geq g_T; \\ 0, & g < g_T, \end{cases} \tag{9.23}$$

where g_T is the cutoff value of the channel power gain. Apparently, when the channel gain is larger than or equal to g_T, the received SNR will be always equal to γ_t. Meanwhile, the system will enter outage when g is less than g_T. Such power adaptation policy can avoid the high-power transmission of full channel inversion when the channel gain is small. The cutoff value g_T can be determined from the desired outage requirement, i.e.,

$$P_{\text{out}} = \Pr[g < g_T] \leq \varepsilon. \tag{9.24}$$

Again, the average transmit power constraint should be satisfied, which implies

$$\int_{g_T}^{\infty} P_t(g)p_g(g)dg \leq \overline{P_t}.$$

(9.25)

The constant received SNR during transmission for given g_T can be shown to be upper bounded by

$$\gamma_t \leq \frac{\overline{P_t}}{N \int_{g_T}^{\infty} (1/g)p_g(g)dg} = \frac{\overline{P_t}}{N \mathbf{E}_{g_T}[1/g]},$$

(9.26)

where $\mathbf{E}_{g_T}[\cdot]$ denotes the statistical averaging over the range of $[g_T, \infty)$.

If, in certain application, the target received SNR γ_t is first specified based on a certain BER requirement, the cutoff value g_T can be numerically calculated from the following equation:

$$\int_{g_T}^{\infty} (1/g)p_g(g)dg = \frac{\gamma_t \cdot N}{\overline{P_t}}.$$

(9.27)

The outage probability of the system can then be calculated as $\Pr[g < g_T]$. Intuitively, for the same fading channel, larger target SNR γ_t will lead to larger outage probability. Therefore, there is a trade-off between outage and reliability during transmission for truncated channel inversion.

Figure 9.7(b) illustrates the operation of truncated channel inversion power adaptation strategy. In particular, the channel power gain variation over time is plotted together with the corresponding instantaneous transmit power. The transmit power becomes zero when the channel power gain is below the cutoff value g_T, leading to zero received SNR over the same time period.

Example: Truncated channel inversion

Consider the power adaptive transmission over a special flat fading channel (same as the one in the previous example for easy comparison). The channel power gain g can take three possible values: $g_1 = 0.001$, $g_2 = 0.01$, and $g_3 = 0.05$ with probability 0.1, 0.6, and 0.3, respectively. The average transmission power $\overline{P_t}$ is 10 mW, the channel bandwidth 30 kHz, and the noise spectrum density 10^{-8} W/Hz. Determine maximum constant received SNR that the system can maintain at the receiver with truncated channel inversion power adaptation strategy with

1. $g_T = 0.003$ and
2. $g_T = 0.03$.

Specify the corresponding power adaptation scheme and the corresponding outage probability for each case.

Solutions:

1. When $g_T = 0.003$, $\mathbf{E}_{g_T}[1/g]$ can be determined as

   $$\mathbf{E}_{g_T}[1/g] = 0.6/0.01 + 0.3/0.05 = 66.$$

   (9.28)

The maximum target SNR under the average transmit power constraint is

$$\gamma_t = \frac{10^{-2}}{30 \times 10^3 \times 10^{-8} \times 66} = 0.51, \tag{9.29}$$

which is larger than that of full channel inversion. Accordingly, the power adaptation policy to maintain such maximum constant SNR is

$$P_t(g) = \begin{cases} 0, & g = g_1; \\ \frac{\gamma_t B_s N_0}{g_2}, & g = g_2; \\ \frac{\gamma_t B_s N_0}{g_3}, & g = g_3. \end{cases} \tag{9.30}$$

$$= \begin{cases} 0, & g = 0.001; \\ 15.2, & g = 0.01; \\ 3.06, & g = 0.05, \end{cases}$$

which leads to an outage probability of 0.1.

2. When $g_T = 0.03$, $\mathbf{E}_{g_T}[1/g]$ becomes $0.3/0.05 = 6$, which leads to $\gamma_t = 5.6$. The power adaptation policy becomes

$$P_t(g) = \begin{cases} 0, & g = 0.001; \\ 0, & g = 0.01; \\ 33.6, & g = 0.05. \end{cases} \tag{9.31}$$

The corresponding outage probability is determined as $0.1 + 0.6 = 0.7$. Clearly, the truncated channel inversion power adaptation policy leads to better performance during transmission at the cost of a certain outage probability. Different cutoff value g_T leads to different tradeoffs between outage probability and received SNR.

9.4   Joint power and rate adaptation

The transmission rate and power of wireless systems can be adapted jointly with the fading channel condition. In this section, we present a joint design of continuous power and discrete rate adaptation.

Let us assume that the transmitter can adaptively transmit with N different transmission schemes, with corresponding transmission rates R_i, $i = 1, 2, \ldots, N$. The instantaneous BER of these transmission schemes are function of the instantaneous SNR, denoted by $P_{b,i}(\gamma)$. The objective of joint power and rate adaptation is to achieve the highest average transmission rate, or equivalently spectral efficiency, while ensuring the instantaneous BER is no more than a target value of BER_0. As such, both the transmission schemes and transmit power should be chosen based on the channel condition. Specifically, the ith transmission scheme with rate R_i and transmit power $P_t(g)$ should be used if

$$P_{b,i}^{-1}(\text{BER}_0) \le \gamma = \frac{P_t(g) \cdot g}{N} < P_{b,i+1}^{-1}(\text{BER}_0), \tag{9.32}$$

where $P_{b,i}^{-1}(\cdot)$ denotes the inverse function of $P_{b,i}(\cdot)$. Therefore, the rate adaptation policy is determined for target BER_0 and given power adaptation policy $P_t(g)$. For notation conciseness, we define $\gamma_{T,i} = P_{b,i}^{-1}(\text{BER}_0)$, for $i = 1, 2, \ldots, N$ while noting that $\gamma_{T,N+1} = \infty$.

The optimal power and discrete rate adaptation strategy can be determined by maximizing

$$\eta = \sum_{i=1}^{N} R_i \int_{\gamma_{T,i} N / P_{t,i}(g)}^{\gamma_{T,i+1} N / P_{t,i}(g)} p_g(g) dg, \tag{9.33}$$

over all power adaptation policy $P_{t,i}(t)$, $i = 1, 2, \ldots, N$, subject to the average transmit power constraint

$$\sum_{i=1}^{N} \int_{\gamma_{T,i} N / P_{t,i}(g)}^{\gamma_{T,i+1} N / P_{t,i+1}(g)} P_{t,i}(g) p_g(g) dg \le \overline{P_t}. \tag{9.34}$$

Here, we assume that different power adaptation policies may apply for different transmission schemes. Unfortunately, directly solving the above optimization problem for optimal power adaptation policy $P_{t,i}(g)$ is very difficult, if not impossible, since $P_{t,i}(g)$ is involved in the integration limits of both the objective function and the constraint.

A suboptimal solution is to apply truncated channel inversion policy for each transmission scheme to satisfy the target BER requirement. Specifically, when transmission scheme with rate R_i is used, the transmission power is set as $P_{t,i}(g) = \gamma_{T,i} N / g$. The optimization problem simplifies to

$$\max_{g_{T_i}, i=1,2,\ldots,N} \sum_{i=1}^{N} R_i \int_{g_{T_i}}^{g_{T_{i+1}}} p_g(g) dg, \tag{9.35}$$

subject to the constraint

$$\sum_{i=1}^{N} \int_{g_{T_i}}^{g_{T_{i+1}}} \frac{\gamma_{T,i} N}{g} p_g(g) dg \le \overline{P_t}. \tag{9.36}$$

The optimal threshold levels g_{T_i} can be solved using the Lagrangian method as follows. The Lagrangian is formulated as

$$J = \sum_{i=1}^{N} R_i \int_{g_{T_i}}^{g_{T_{i+1}}} p_g(g) dg - \lambda \left(\sum_{i=1}^{N} \int_{g_{T_i}}^{g_{T_{i+1}}} \frac{\gamma_{T,i} N}{g} p_g(g) dg - \overline{P_t} \right). \tag{9.37}$$

Taking partial derivative of J with respect to g_{T_i}, $i = 1, 2, \ldots, N$, and setting the results equal to zero, we arrive at N equations with N unknowns, given by

$$\frac{dJ}{dg_{T_1}} = -R_1 - \lambda \left(-\frac{\gamma_{T,1} N}{g_{T_1}} \right) = 0, \tag{9.38}$$

and

$$\frac{dJ}{dg_{T_i}} = (R_{i-1} - R_i) - \lambda \left(\frac{\gamma_{T,i-1} N}{g_{T_i}} - \frac{\gamma_{T,i} N}{g_{T_i}} \right) = 0, \quad i = 2, \ldots, N. \tag{9.39}$$

We can determine the optimal threshold levels for channel power gain g as

$$g_{T_1} = \frac{R_1}{\lambda N \gamma_{T,1}}; \quad g_{T_i} = \frac{R_i - R_{i-1}}{\lambda N (\gamma_{T,i-1} - \gamma_{T,i-1})}, \quad i = 2, \ldots, N, \tag{9.40}$$

where the value of λ is determined by the average power constraint given by

$$\sum_{i=1}^{N} \int_{g_{T_i}}^{g_{T_{i+1}}} \frac{\gamma_{T,i} N}{g} p_g(g) dg = \overline{P_t}. \tag{9.41}$$

9.5 Implementation issues

Channel adaptive transmission is an effective solution to improve the reliability and spectrum efficiency of digital wireless communication systems. The basic premise is that the transmitter and the receiver can be adaptively configured based on the prevailing fading channel condition. It has been shown in previous chapter that with perfect channel state information (CSI) at the transmitter, there exists an optimal adaptive transmission scheme, involving continuous rate and power adaptation, which can achieve the Shannon capacity over fading channels. However, providing perfect CSI at the transmitter is a very challenging task in reality, even for the today's most advanced wireless systems. In addition, implementing continuous rate adaptation will entail a prohibitively high transceiver complexity. As a result, while acknowledging a certain performance gap compared to the optimal scheme, most current wireless standards adopt adaptive transmission schemes employing discrete rate adaptation and suboptimal power adaptation strategies discussed in this chapter.

The availability of the channel knowledge is a fundamental requirement of adaptive transmission techniques. With discrete rate adaptation, the transmitter needs to know the transmission rate that the current channel condition can support. Such information is typically obtained through receiver feedback. In particular, the receiver estimates the instantaneous received SNR and compares it with predetermined threshold levels to select the suitable transmission rate. Then, the receiver will feedback the rate selection result to the transmitter. As such, feedback mechanism is critical to rate adaptation as both transmitter and receiver should be configured to the same MCS scheme. With power adaptation, the transmitter can also estimate the downlink channel power gain by exploring the channel reciprocity. Specifically, if the uplink channel gain is highly correlated with the downlink channel gain, the transmitter can predict the downlink channel gain based on its received signal and use the prediction result to set the transmission power level.

The transmitter and the receiver must be reconfigurable to implement rate adaptive transmission. Both the modulation and coding schemes may need to be changed with the prevailing channel condition. The rate of channel variation dictates the frequency of such reconfiguration. As such, adaptive transmission systems are more suitable for slow fading channels. In particular, at the start of a channel coherence time, the system performs channel estimation, rate selection, and transceiver reconfiguration. The resulting configuration will be used for the remaining of the channel coherence time until updated channel estimation becomes available. Over fast fading environment, the performance benefit of adaptive transmission system will be greatly compromised due to such overhead for reconfiguration. An alternative approach for the fast fading scenario is to track and adapt to the shadowing variation and leave the effect of fading to other fading mitigation techniques.

9.6   Further readings

Early investigation on adaptive transmission dated back to the 1970s [1,2]. References [3,4] present early adoption of adaptive transmission in wireless standards. Further discussion about joint rate and power adaptive transmission can be found in [5, Chapter 9].

Problems

1. A digital transmitter employing M-QAM modulation scheme is operating over a fading channel with an instantaneous received SNR of 15 dB. The coding rate is assumed to be 3/4, and the symbol rate is 2 Msps. Determine the information data rate and the approximate BER when (i) $M = 16$, (ii) $M = 64$, and (iii) $M = 256$.

2. Consider an adaptive modulation system using three modulation schemes, QPSK, 8-PSK, and 16-PSK. The BER of MPSK with Gray coding can be approximated calculated as

$$P_b(\gamma) \approx \frac{2}{\log_2 M} Q\left(\sqrt{2\gamma}\sin(\pi/M)\right). \tag{9.42}$$

   The modulation schemes are chosen adaptively according to the instantaneous channel condition to satisfy the instantaneous BER requirement of 10^{-4}. Determine the mode of operation of the adaptation modulation scheme.

3. Consider an adaptive modulation system that uses the following three different modulation schemes: QPSK, 16-QAM, and 64-QAM. The modulation schemes are adaptively selected to satisfy the instantaneous BER requirement of 10^{-3}.

Assuming the system is operating over a Rayleigh fading channel with an average SNR of 18 dB,

   (i)   Determine the average spectral efficiency of the system in bits/symbol.

   (ii)   What if the average SNR changes to 8 dB? Compare your result with conventional system that uses QPSK all the time.

4. Repeat your calculation of average spectral efficiency in previous question while assuming an MRC diversity combiner with $L = 3$ independent and identically faded antenna branches is implemented at the receiver. You will find the CDF of the combined SNR with MRC over i.i.d. branches given in Chapter 5 useful.

5. A power adaptive transmission system is operating over a Nakagami fading channel. The average channel power gain is 0.1, and the Nakagami m parameter is equal to 2. The average transmit power over noise power ratio $\overline{P_t}/N = 15$ dB. Numerically calculate and compare the maximum constant SNR that the system can maintain when (i) full channel inversion and (ii) truncated channel inversion with outage probability at most 0.05 is applied.

6. The channel power gain of a special flat fading channel takes four possible values: $g_1 = 0.001, g_2 = 0.01, g_3 = 0.03$, and $g_4 = 0.05$ with probability 0.1, 0.3, 0.2, and 0.4, respectively. The average transmission power $\overline{P_t}$ is 20 mW, the noise power 1 mW. Determine maximum constant received SNR that the system can maintain at the receiver with (i) full channel inversion and (ii) truncated channel inversion with outage probability requirement of 0.2. Specify the corresponding power adaptation schemes.

7. Let consider the digital wireless transmission over Rayleigh flat fading channel. The average channel power gain is $\overline{g} = 5$ dB. The average transmit power over the noise power ratio $\overline{P_t}/N$ is 15 dB. The system adopt truncated channel inversion to satisfy the target BER of 10^{-3} during transmission with BPSK modulation. Determine the maximum target SNR that the system can maintain and the corresponding outage probability. What if the target BER is changed to 10^{-5}?

8. An adaptive transmission system uses QPSK and 8-PSK in combination of block codes with rate 1/2 and 1/3. Assume that the rate 1/2 code will provide 3 dB improvement in instantaneous BER, and rate 1/3 code provides 5 dB improvement. Determine the adaptive transmission scheme for the target BER of 10^{-5}, i.e., find the coding and modulation scheme combination for each possible value of received SNR, to achieve the maximum data rate while satisfy the target BER requirement. Determine the feedback load of the system in terms of number of bits per channel coherence time T_c.

Bibliography

[1]   J. F. Hayes, "Adaptive feedback communications", *IEEE Trans. Commun. Tech.*, Vol. 16, pp. 29–34, February 1968.

[2]   J. K. Cavers, "Variable-rate transmission for Rayleigh fading channels", *IEEE Trans. Commun.*, Vol. 20, pp. 15–22, February 1972.

[3]   A. Furuskar, S. Mazur, F. Muller, and H. Olofsson, "EDGE: Enhanced data rates for GSM and TDMA/136 evolution", *IEEE Wireless Commun. Mag.*, Vol. 6, pp. 56–66, June 1999.

[4]   A. Ghosh, L. Jalloul, B. Love, M. Cudak, and B. Classon, "Air-interface for 1XTREME/1xEV-DV", *Proc. 53rd IEEE Veh. Tech. Conf.*, pp. 2474–2478, May 2001.

[5]   A. Goldsmith, *Wireless Communications*, New York, NY: Cambridge University Press, 2005.

Chapter 10
MIMO transmission

In previous chapter, we considered the scenarios where multiple transmit antennas or multiple receive antennas were deployed and demonstrated the resulting diversity benefit. We now consider the scenario where both the transmitter and the receiver are equipped with multiple antennas. Multiple-antenna transmission and reception techniques, also known as multiple-input-multiple-output (MIMO) technology, can considerably improve the performance and efficiency of wireless communication systems. MIMO transmission has the capability to achieve spatial multiplexing gain as well as diversity gain. Significant research efforts have been devoted to the development of MIMO transmission schemes and their application in real-world wireless communication systems.

In this chapter, we study the fundamental principles of MIMO transmission and illustrate its performance advantages. We first introduce the flat fading MIMO channel model. Then, we present several MIMO transmission strategies exploring the diversity benefit inherent to MIMO channel. After that, we investigate the capacity potential of MIMO channel and study MIMO transmission strategies extracting the spatial multiplexing gains. The chapter is concluded with a characterization of the diversity-multiplexing trade-off.

10.1 MIMO channel model

Let us consider a point-to-point link where the transmitter has M_t antennas, and the receiver has M_r antennas, as shown in Figure 10.1. We assume that the channel for

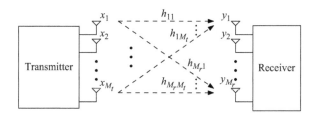

Figure 10.1 Generic MIMO channel model

the jth transmit antenna to the ith receiver antenna introduces frequency flat fading, with complex channel gain denoted by h_{ij}, $i = 1, 2, \ldots, M_r$, $j = 1, 2, \ldots, M_t$. Let x_j, $j = 1, 2, \ldots, M_t$ denote the transmitted symbol from the ith transmit antenna over a symbol period. The received symbol at the ith receive antenna is given by

$$y_i = h_{i1}x_1 + h_{i2}x_2 + \cdots + h_{iM_t}x_{M_t} + n_i, \quad i = 1, 2, \ldots, M_r, \tag{10.1}$$

where n_i's are the samples of independent white Gaussian noise with zero mean and variance $\sigma^2 = N_0$. If we arrange the received symbols on M_r receive antennas in a vector form, then the input/output relationship of the MIMO channel is given by

$$\begin{bmatrix} y_1 \\ y_2 \\ \vdots \\ y_{M_r} \end{bmatrix} = \begin{bmatrix} h_{11} & h_{12} & \cdots & h_{1M_t} \\ h_{21} & h_{22} & \cdots & h_{2M_t} \\ \vdots & \vdots & \ddots & \vdots \\ h_{M_r 1} & h_{M_r 2} & \cdots & h_{M_r M_t} \end{bmatrix} \begin{bmatrix} x_1 \\ x_2 \\ \vdots \\ x_{M_t} \end{bmatrix} + \begin{bmatrix} n_1 \\ n_2 \\ \vdots \\ n_{M_r} \end{bmatrix} \tag{10.2}$$

or simply in matrix notation

$$\mathbf{y} = \mathbf{Hx} + \mathbf{n}, \tag{10.3}$$

where $\mathbf{x}$ denote the column vector of transmitted symbols, $\mathbf{y}$ the vector of received symbols, $\mathbf{n}$ the vector of noise samples, and $\mathbf{H}$ is the $M_r \times M_t$ matrix of channel gains h_{ij}. Let P_{tot} denote the maximum total transmission power from all M_t antenna. Then, we have the following total transmit power constraint

$$\sum_{j=1}^{M_t} \mathbf{E}[x_i x_i^*] \leq P_{\text{tot}}, \tag{10.4}$$

where $\mathbf{E}[\cdot]$ denotes statistical expectation operation. Alternatively, this constraint can be given in terms of the covariance matrix of transmitted symbols $\mathbf{R_x} = \mathbf{E}[\mathbf{xx}^H]$ as

$$\text{Tr}(\mathbf{R_x}) \leq P_{\text{tot}}, \tag{10.5}$$

where $\text{Tr}(\cdot)$ returns the trace of a matrix.

The multiple antennas at the transmitter and the receiver can be exploited to improve the performance and spectral efficiency of wireless transmission. The amount of performance improvement depends on several factors, including the characteristics of channel gain matrix $\mathbf{H}$ and its availability at the transmitter and the receiver. The entries of $\mathbf{H}$ are commonly assumed to be independent and identically distributed (i.i.d.) zero-mean complex Gaussian random variables. Such assumptions are normally valid in rich scattering environment when the antennas are separated further enough. The knowledge of the channel matrix at the receiver is usually obtained through the channel estimation process. Meanwhile, the transmitter will rely on receiver feedback or channel reciprocal property to know $\mathbf{H}$. Typically, it is more challenging for the transmitter to acquire the complete knowledge of channel matrix $\mathbf{H}$.

10.2 Diversity over MIMO channels

Multiple antennas at the transmitter and the receiver can be explored for diversity benefit. If a certain knowledge about the channel matrix **H** can be made available to the transmitter, i.e., channel state information at the transmitter (CSIT) scenario, we can apply conventional diversity transmission/reception schemes. If the channel knowledge is only available at the receiver, space-time coded transmission combined with diversity reception can be employed to extract maximum diversity gain.

10.2.1 Channel knowledge available at transmitter

When the CSI is available at the transmitter a straightforward strategy to explore diversity benefit is to apply selection combining over all transmit and receive antenna pairs. Specifically, the receiver estimates and compares the channel power gains $|h_{ij}|^2$, for all i and j. On the basis of the comparison result, the receiver selects the antenna pair that achieves the largest power gain, i.e., transmit antenna j^* and receive antenna i^* such that

$$|h_{i^*j^*}|^2 = \max_{i,j}|h_{ij}|^2. \tag{10.6}$$

The receiver will then feedback the index of the chosen transmit antenna j^*, which incurs a feedback load of $\log_2 M_t$ per channel coherence time, and use the corresponding receive antenna for reception. Since there are $M_t M_r$ antenna pairs to choose from, such full selection strategy essentially implements an $M_t M_r$-branch selection combining scheme. Note that with conventional transmit or receive antenna diversity solution, we will need to deploy a total of $M_t M_r$ antennas to achieve the same diversity benefit. With MIMO implementation, we only need to deploy $M_t + M_r$ antennas. Such hardware complexity saving is more significant when both M_t and M_r are large. The structure of MIMO diversity system with full selection is shown in Figure 10.2(a).

An alternative strategy is to perform antenna selection at the transmitter and maximum ratio combining (MRC) at the receiver. In particular, the receiver estimates the channel matrix **H** and use it to calculate the combined channel power gain corresponding to each transmit antenna when MRC is applied at the receiver, which is given for jth transmit antenna by $\sum_{i=1}^{M_r} |h_{ij}|^2$. After comparison, the receiver selects the transmit antenna that leads to the largest combined channel power gain by feeding its index, denoted by j^*, back to the transmitter. The receiver then applies MRC with the combining weight for ith antenna proportional to $h_{ij^*}^*$. This so-called transmit antenna selection with receiver MRC (TAS/MRC) strategy entails the same feedback load as full selection combining scheme above but achieves better performance as intuitively expected. The TAS/MRC diversity system is illustrated in Figure 10.2(b).

The performance of MIMO transmission can be further improved when all M_t transmit antenna are used for transmission. We now derive the optimal linear transmission and reception scheme in terms of maximizing the diversity benefit of MIMO channels. The structure of the transmitter and the receiver are shown in Figure 10.2(c). Specifically, over each symbol period, a data symbol s is transmitted from all M_t antennas after being weighted by vector $\mathbf{v} = \{v_1, v_2, \ldots, v_{M_t}\}^T$. The transmitted symbol from antenna j is $x_j = v_j s$, $j = 1, 2, \ldots, M_t$. Vector $\mathbf{v}$ should be normalized to

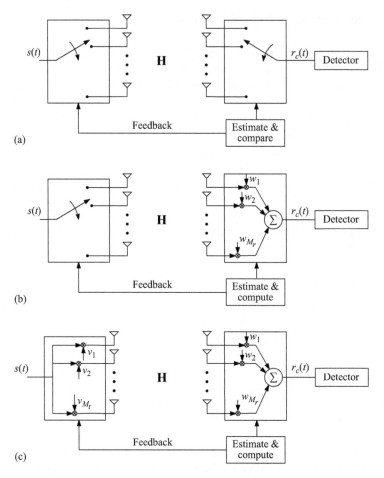

Figure 10.2   Diversity over MIMO channels with CSI at the transmitter: (a) MIMO with antenna selection, (b) MIMO with TAS/MRC, and (c) MIMO with MRT/MRC

satisfy the total transmit power constraint, i.e., $\|\mathbf{v}\| \leq 1$. The receiver will linearly combine the received symbols over all M_r antennas with normalized weighting vector $\mathbf{w} = \{w_1, w_2, \ldots, w_{M_r}\}^T$, $\|\mathbf{w}\| = 1$. The combined signal at the receiver is given by

$$r = w_1 y_1 + w_2 y_2 + \cdots + w_{M_r} y_{M_r} = \mathbf{w}^T \mathbf{y}. \tag{10.7}$$

Applying (10.3) while noting $\mathbf{x} = \mathbf{v}s$, we have

$$r = \mathbf{w}^T \mathbf{H} \mathbf{v} s + \tilde{n}, \tag{10.8}$$

where $\tilde{n} = \mathbf{w}^T \mathbf{n}$. Since $\mathbf{w}$ is normalized, $\tilde{n}$ can be shown to be Gaussian distributed with zero mean and variance σ^2.

The maximum diversity benefit can be achieved by optimally designing vectors **v** and **w** to maximize the SNR of combined signal, given by

$$\gamma = \frac{|\mathbf{w}^T\mathbf{H}\mathbf{v}|^2 E_s}{\sigma^2}. \tag{10.9}$$

We arrive at the following optimization problem

$$\max_{\mathbf{v},\mathbf{w}} \quad \frac{|\mathbf{w}^T\mathbf{H}\mathbf{v}|^2 E_s}{\sigma^2} \tag{10.10}$$

s.t. $\|\mathbf{v}\| \leq 1, \|\mathbf{w}\| = 1$.

The solution to this problem can be easily obtained from matrix theory. Specifically, it has been established that $|\mathbf{w}^T\mathbf{H}\mathbf{v}|^2$ reaches its maximum value, which is the largest singular value of **H**, denoted by λ_{max}, when **v** is chosen to be the corresponding right singular vector, and **w** is the complex conjugate of the corresponding left singular vector **u**, i.e., $\mathbf{w} = \mathbf{u}^*$. The resulting diversity scheme for MIMO channel is also referred to as *MIMO beamforming*. It can achieve the largest received SNR, given by

$$\gamma = \frac{\lambda_{max} E_s}{\sigma^2}, \tag{10.11}$$

with linear processing at the transmitter and the receiver. Meanwhile, the transmitter needs to know the exact right singular vector corresponding to λ_{max}. Should the vector $\mathbf{u}^*$ be fed back from the receiver, the required feedback load is M_t complex numbers per channel coherence time.

Example: MIMO diversity

Consider a MIMO wireless channel with four transmit antennas and three receive antennas. The instantaneous channel gain matrix is given by

$$\mathbf{H} = \begin{bmatrix} 0.35 + j0.23 & 0.45 - j0.14 & 0.2 + j0.03 & 0.4 - j0.42 \\ 0.15 - j0.24 & 0.4 + j0.11 & 0.1 - j0.13 & 0.2 + j0.05 \\ 0.05 + j0.12 & 0.15 + j0.17 & 0.3 - j0.24 & 0.3 - j0.35 \end{bmatrix}. \tag{10.12}$$

Assuming the symbol energy over noise power ratio E_s/σ^2 is equal to 10 dB, determine the value of the instantaneous combined SNR at the receiver with

1. full selection diversity,
2. TAS/MRC diversity, and
3. MRT/MRC diversity.

Solutions:

1. With full selection, the maximum channel power gain is reached when the fourth transmit antenna and the first receive antenna are selected for transmission, i.e.,

$$|h_{i^*j^*}|^2 = \max_{i,j} |h_{14}|^2. \tag{10.13}$$

The resulting received SNR is calculated as $|0.4 - j0.42|^2 \times 10 = 3.36 = 5.2$ dB.

2. The combined SNR at the receiver corresponding to four transmit antennas
    are determined as 2.7, 4.45, 2.15, and 5.91, respectively. As such, the fourth
    transmit antenna should be used for transmission with TAS/MRC scheme,
    which leads to the received SNR of 5.91 = 7.7 dB.
3. With MRT/MRC scheme, the combined SNR depends upon the largest
    singular value of **H**, λ_{max}. Applying singular value decomposition (SVD)
    operation (using svd() command in MATLAB®) to **H**, λ_{max} can be deter-
    mined as 1.07. As such, the instantaneous combined SNR is calculated as
    $1.07 \times 10 = 10.7 = 10.3$ dB.

10.2.2   Channel knowledge unavailable at transmitter

When the transmitter has absolute no knowledge about the channel matrix **H** (CSI at
the receiver only scenario), we can employ space-time coded transmission combined
with reception diversity to extract diversity benefit. To illustrate further, let consider a
MIMO channel with $M_t = 2$ transmit antennas and M_r receive antennas, as shown in
Figure 10.3. For dual transmit antenna case, the Alamouti scheme discussed in earlier
chapter can be used for full rate diversity transmission. The received signal over two
consecutive symbol periods at the ith receive antenna are given by

$$r_{i1} = h_{i1}s_1 + h_{i2}s_2 + n_{i1},$$
$$r_{i2} = h_{i1}(-s_2^*) + h_{i2}s_1^* + n_{i2}, \quad i = 1, 2, \ldots, M_r, \tag{10.14}$$

where s_1 and s_2 are the transmitted data symbols, n_{i1} and n_{i2} are the noise sample at
the ith antenna over the first and second symbol periods, respectively. We assume that
the noise samples at different antennas and/or symbol periods are independent with
zero mean and same variance N_0. After linear processing at the receiver, the effective
received symbol for s_1 at the ith antenna is given by

$$z_{i1} = (|h_{i1}|^2 + |h_{i2}|^2)s_1 + \tilde{n}_{i1}, \quad i = 1, 2, \ldots, M_r, \tag{10.15}$$

where $\tilde{n}_{i1} = h_{i1}^* n_{i1} + h_{i2}n_{i2}^*$ is additive noise, following a complex Gaussian distribu-
tion with zero mean and variance $(|h_{i1}|^2 + |h_{21}|^2)N_0$. The effective received symbol
for s_2 at the ith antenna can similarly calculated, so can those at other receive antennas.

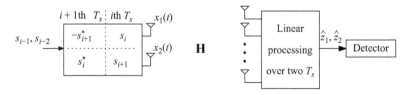

Figure 10.3   Diversity over MIMO channels without CSI at the transmitter

The receiver will then add the effective received symbols at M_r receive antennas together to generate the decision statistics for s_1. Specifically, the decision statistics for s_1 is calculated as

$$\hat{z}_1 = \sum_{i=1}^{M_r} z_{i1} = \sum_{i=1}^{M_r} (|h_{i1}|^2 + |h_{i2}|^2)s_1 + \sum_{i=1}^{M_r} \tilde{n}_{i1}, \tag{10.16}$$

and that for s_2 as

$$\hat{z}_2 = \sum_{i=1}^{M_r} z_{i2} = \sum_{i=1}^{M_r} (|h_{i1}|^2 + |h_{i2}|^2)s_2 + \sum_{i=1}^{M_r} \tilde{n}_{i2}. \tag{10.17}$$

The symbol energy is $E_s/2$ due to the double transmission of each symbol. The power of the noise term in the decision statistics can be shown to be equal to $\sum_{i=1}^{M_r} (|h_{i1}|^2 + |h_{i2}|^2)N_0$. Therefore, the SNR of the decision statistics can be shown to be given by

$$\gamma_c = \sum_{i=1}^{M_r} (|h_{i1}|^2 + |h_{i2}|^2)E_s/2N_0 = \|\mathbf{H}\|_F^2 E_s/2N_0, \tag{10.18}$$

where $\|\mathbf{H}\|_F = \sqrt{\sum_{i=1}^{M_r} (|h_{i1}|^2 + |h_{i2}|^2)}$ denotes the Frobenius norm of the channel matrix $\mathbf{H}$. Therefore, the overall system is equivalent to a $2M_r$-branch MRC receiver except for a 3-dB loss. When there are more than two transmit antennas, we can apply generalized orthogonal space-time codes at the transmitter together with receive diversity to extract full diversity benefit of MIMO channel at the cost of a certain rate loss.

The performance improvement through diversity combining, in terms of the reduction in average error rate and outage probability, can be characterized by the diversity order of the system. Diversity order is defined as the decreasing rate of average error rate of the system as the average received SNR increases. If the average error rate of a diversity system can be written as the function of the average SNR $\bar{\gamma}$ as $\bar{P}_e = C\bar{\gamma}^{d_o}$, where C is a constant independent of $\bar{\gamma}$, then the system achieves a diversity order of d_o. In general, the diversity order is calculated based on the asymptotic behavior of average error rate as

$$d_o = -\lim_{\bar{\gamma} \to \infty} \frac{\log \bar{P}_e(\bar{\gamma})}{\log \bar{\gamma}}, \tag{10.19}$$

where the logarithm can be calculated in any base. Typically, d_o can be observed from average error rate curve, plotted in logarithm as function of the average SNR in dB scale, as the negative slope over high SNR region. Figure 10.4 illustrates the concept of diversity order.

The maximum diversity order of a system with L diversity branches is L. If the diversity order of a system is equal to the maximum L, then the system is claimed to achieve full diversity order. It has been shown that MRC achieves full diversity order in diversity reception system. Space-time coded transmission also achieves full diversity order over transmit diversity system without the requirement of channel state

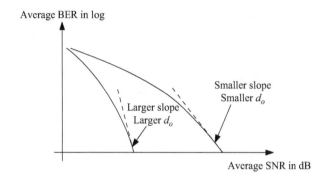

Figure 10.4   Diversity order from average error rate curve

information at the transmitter. Finally, space-time coded transmission combined with MRC reception achieves a full diversity order of $M_t M_r$ over MIMO channels.

10.3   Multiplexing over MIMO channels

Multiple antennas at the transmitter and the receiver can also be exploited for spatial multiplexing gain. The spectral efficiency of wireless transmission can be enhanced by exploring the spatial degree of freedom inherent to multiple antennas. When CSI is available at the transmitter, the spatially multiplexing benefit can be more efficiently exploited. For the CSI at the receiver only scenario, various transmission strategies, such as space-time coded transmission and Bell Labs layered space-time (BLAST) structures, can apply.

10.3.1   Parallel decomposition

The most informative way to demonstrate the spatial multiplexing capability of MIMO channel is parallel decomposition. Specifically, the MIMO channel can be decomposed into multiple parallel channels as following. The MIMO channel matrix **H**, after applying SVD, can be rewritten as

$$\mathbf{H}_{M_r \times M_t} = \mathbf{U}_{M_r \times M_r} \mathbf{\Sigma}_{M_r \times M_t} \mathbf{V}^H_{M_t \times M_t}, \tag{10.20}$$

where Σ is a diagonal matrix with diagonal entries being the singular values of **H**, denoted by σ_i, **U**, and **V** are the left and right singular matrices, respectively, i.e., $\mathbf{U}^H \mathbf{U} = I_{M_r}$ and $\mathbf{V}^H \mathbf{V} = I_{M_t}$. Note that σ_i^2 is the ith largest eigenvalue of matrix $\mathbf{H}\mathbf{H}^H$. It is not difficult to see that matrix $\mathbf{H}_{M_r \times M_t}$ will have at most $\min[M_t, M_r]$ nonzero singular values. Let $R_\mathbf{H}$ denote the number of nonzero singular values, also known as *rank* of matrix **H**, with $R_\mathbf{H} \in [1, \min\{M_t, M_r\}]$. Then, the MIMO channel can be transformed into $R_\mathbf{H}$ parallel independent channels as follows.

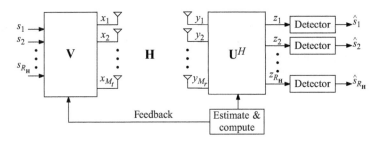

Figure 10.5 Parallel decomposition of MIMO channels

We first generate the transmitted symbols from M_t antennas over a symbol period $\mathbf{x}$ by applying precoding operation to data symbols, denoted by $\mathbf{s}$, as

$$\mathbf{x} = \mathbf{V}\mathbf{s}. \tag{10.21}$$

Here, the right singular matrix $\mathbf{V}$ is used as the precoding matrix. The received symbols over the MIMO channel can be written as

$$\mathbf{y} = \mathbf{H}\mathbf{x} + \mathbf{n} = \mathbf{H}\mathbf{V}\mathbf{s} + \mathbf{n}. \tag{10.22}$$

The receiver will then apply the shaping operation with shaping matrix $\mathbf{U}$ as $\mathbf{z} = \mathbf{U}^H\mathbf{y}$. The output symbols can be shown to be given by

$$\mathbf{z} = \Sigma_{M_r \times M_t}\mathbf{s} + \mathbf{U}^H\mathbf{n}. \tag{10.23}$$

Note that Σ is a diagonal matrix with diagonal entries σ_is. The ith entry of $\mathbf{z}$, $1 \leq i \leq R_{\mathbf{H}}$, can be written as

$$z_i = \sigma_i s_i + \tilde{n}_i, \tag{10.24}$$

where $\tilde{n}_i$ is the ith entry of the noise vector $\mathbf{U}^H\mathbf{n}$. It can be shown, while noting that $\mathbf{U}$ is a unitary matrix, that $\tilde{n}_i$ is still white Gaussian with zero mean and variance N_0. As such, we can detect data symbol s_i based on z_i in presence of channel scaling factor σ_i and additive noise $\tilde{n}_i$. As many as $R_{\mathbf{H}}$ data symbols can be transmitted in parallel over the MIMO channel at the same time over the same frequency without mutual interference. Effectively, the data rate of the system can be increased by $R_{\mathbf{H}}$ times, leading to a multiplexing gain factor of $R_{\mathbf{H}}$. The process of MIMO channel decomposition is illustrated in Figure 10.5. Note that the transmitter needs to have the complete knowledge of precoding matrix $\mathbf{V}$ to perform such parallel transmission.

Example: Parallel decomposition

Consider a MIMO wireless channel with four transmit antennas and three receive antennas. The instantaneous channel gain matrix is given by

$$\mathbf{H} = \begin{bmatrix} 0.35 + j0.23 & 0.45 - j0.14 & 0.2 + j0.03 & 0.4 - j0.42 \\ 0.15 - j0.24 & 0.4 + j0.11 & 0.1 - j0.13 & 0.2 + j0.05 \\ 0.05 + j0.12 & 0.15 + j0.17 & 0.3 - j0.24 & 0.3 - j0.35 \end{bmatrix}. \tag{10.25}$$

Determine the precoding matrix that can be used at the transmitter to decompose the channel into parallel independent channels. What are the channel power gain of the decomposed channels?

Solutions: Applying SVD operation, we determine the precoding matrix **V** as

$$
\mathbf{V} = \begin{bmatrix}
0.312 - j0.14 & 0.29 + j0.50 & -0.59 + j0.19 & 0.38 + j0.12 \\
0.54 + j0.02 & 0.42 - j0.45 & -0.26 - j0.26 & -0.44 \\
0.30 + j0.16 & -0.06 + j0.34 & 0.28 + j0.57 & -0.54 + j0.28 \\
0.47 + j0.50 & -0.22 - j0.34 & 0.15 + j0.23 & 0.53 - j0.01
\end{bmatrix}. \tag{10.26}
$$

Also from the SVD operation, we obtain Σ matrix as

$$
\Sigma = \begin{bmatrix}
1.07 & 0 & 0 & 0 \\
0 & 0.48 & 0 & 0 \\
0 & 0 & 0.38 & 0
\end{bmatrix}. \tag{10.27}
$$

Therefore, the channel gain of the decomposed channels are 1.07, 0.48, and 0.38.

On the basis of the above parallel decomposition, the instantaneous capacity of MIMO wireless channel can be determined as the sum of the capacity of individual decomposed channels. Let P_i denote the power allocation to the ith decomposed channel, which satisfies a total power constraint of $\sum_i P_i \le P_{\text{tot}}$. The MIMO capacity can be calculated as

$$
C_{\text{inst}} = \sum_{i=1}^{R_{\mathrm{H}}} B \log_2 \left(1 + \frac{\sigma_i^2 P_i}{N_0} \right). \tag{10.28}
$$

If the transmitter does not know σ_i^2s, equal power allocation over decomposed channel should apply, which leads to the channel capacity

$$
C_{\text{equal}} = \sum_{i=1}^{R_{\mathrm{H}}} B \log_2 \left(1 + \frac{\sigma_i^2 P_{\text{tot}}}{R_{\mathrm{H}} N_0} \right). \tag{10.29}
$$

With the knowledge of σ_i at the transmitter, the above instantaneous capacity can be increased by performing optimal transmit power allocation over the decomposed channels. Specifically, we need to solve the following optimization problem

$$
\max_{\{P_i\}_{i=1}^{R_{\mathrm{H}}}} : \sum_{i=1}^{R_{\mathrm{H}}} B \log_2 \left(1 + \frac{\sigma_i^2 P_i}{N_0} \right); \tag{10.30}
$$

$$
\text{s.t.} \quad \sum_i P_i \le P_{\text{tot}}.
$$

Solving the optimization problem using Lagrangian method, we again arrive at the well-known water-filling solution, given by

$$P_i^* = \begin{cases} N_0(1/g_0 - 1/\sigma_i^2), & \sigma_i^2 \geq g_0, \\ 0, & \sigma_i^2 < g_0, \end{cases} \tag{10.31}$$

where g_0 is the cutoff value based on the total power constraint

$$\sum_{i:\sigma_i^2 \geq g_0} P_i = P_{\text{tot}}. \tag{10.32}$$

The resulting instantaneous channel capacity is equal to

$$C_{\text{opt}} = \sum_{i:\sigma_i^2 \geq g_0} B \log_2 \left(\frac{\sigma_i^2}{g_0} \right). \tag{10.33}$$

Example: MIMO channel capacity

Consider a MIMO wireless channel with four transmit antennas and three receive antennas. The instantaneous channel gain matrix is given by

$$\mathbf{H} = \begin{bmatrix} 0.35 + j0.23 & 0.45 - j0.14 & 0.2 + j0.03 & 0.4 - j0.42 \\ 0.15 - j0.24 & 0.4 + j0.11 & 0.1 - j0.13 & 0.2 + j0.05 \\ 0.05 + j0.12 & 0.15 + j0.17 & 0.3 - j0.24 & 0.3 - j0.35 \end{bmatrix}. \tag{10.34}$$

Assuming the transmit power over noise spectrum density ratio P_{tot}/N_0 is equal to 10 dB, determine the normalized instantaneous capacity of the channel C/B with

1. equal power allocation and
2. optimal power allocation.

Solutions:

1. Applying the SVD result from the previous example, the channel power gain of the decomposed channels are $\sigma_1^2 = 1.15$, $\sigma_2^2 = 0.23$, and $\sigma_3^2 = 0.15$. Therefore, we can determine C_{equal}/B as

$$\frac{C_{\text{equal}}}{B} = \sum_{i=1}^{3} \log_2 \left(1 + \frac{10\sigma_i^2}{3} \right) = 2.28 + 0.81 + 0.57 = 3.66 \text{ bps/Hz.}$$

$$\tag{10.35}$$

2. To calculate the capacity with optimal power allocation, we first determine g_0. Let us first assume $g_0 < \sigma_3^2$ and all three channels are used. The cutoff value g_0 can be determined from the power constraint

$$\sum_{i=1}^{3} (1/g - 1/\sigma_i^2) = 10, \tag{10.36}$$

as $g_0 = 0.14 < \sigma_3^2$. There is no conflict. The channel capacity can be calculated as

$$\frac{C_{\text{opt}}}{B} = \sum_{i=1}^{3} \log_2 \left(\frac{\sigma_i^2}{0.14} \right) = 3.04 + 0.72 + 0.1 = 3.77 \text{ bps/Hz.} \qquad (10.37)$$

As the result of multipath fading, h_{ij}s are in general time varying. The ergodic capacity of MIMO channel over fading channels can be obtained by averaging the instantaneous capacity over the fading distribution of channel matrix $\mathbf{H}$ as

$$C_f = \mathbf{E}_{\mathbf{H}}[C_{\text{opt}}(\mathbf{H})], \qquad (10.38)$$

or equivalently, over the distribution of the singular values of $\mathbf{H}$, σ_i, as

$$C_f = \mathbf{E}_{\sigma_i} \left[C_{\text{opt}}(\sigma_i) \right]. \qquad (10.39)$$

The expectation can be solved in some cases while noting that σ_i^2 is the ith largest eigenvalues of the matrix $\mathbf{H}\mathbf{H}^H$.

10.3.2   Spatial multiplexing

The spatial multiplexing gain of MIMO wireless channels can be readily achieved when the CSI is available at both the transmitter and the receiver. Specifically, with the knowledge of the right singular matrix $\mathbf{V}$ at the transmitter and the left singular matrix $\mathbf{U}$ at the receiver, the system can decompose the MIMO channel into $R_{\mathbf{H}}$ parallel channels and transmit independent data streams over each channel. Different decomposed channels may use different coding and modulation schemes to transmit at different information rate with a common symbol rate. Depending on whether the transmitter has the additional knowledge of the singular values σ_is or not, the transmitter may apply either equal power allocation or optimal power allocation, which further enhances the overall transmission rate over these parallel channels.

In practical transmission systems, it is very challenging to make the CSI information perfectly available at the transmitter, especially for MIMO wireless channels. Note that for FDD implementation, equipping $\mathbf{V}$ at the transmitter will involve a feedback load of M_t^2 complex numbers every channel coherence time. When only the receiver has the complete knowledge of the MIMO channel gain matrix $\mathbf{H}$, the spatial multiplexing gain of MIMO channels can still be achieved, at the cost of higher receiver complexity.

10.3.2.1   Uncoded transmission

A straightforward approach is to transmit M_t independent data symbols over M_t transmit antennas in each symbol period. The receiver will then apply joint detection on the received symbols on M_r received antennas, given by

$$\mathbf{y} = \mathbf{H}\mathbf{s} + \mathbf{n}, \qquad (10.40)$$

to recover transmitted data symbols. Intuitively, we expect $M_r \geq M_t$ for decent detection performance. The optimal detection scheme will be the maximum likelihood detection, which detects $\hat{\mathbf{s}}$ as

$$\hat{\mathbf{s}} = \arg\min_{\mathbf{s}} \|\mathbf{y} - \mathbf{H}\mathbf{s}\|^2. \tag{10.41}$$

It can be shown that the average pairwise error probability, i.e., transmitted data symbol vector $\mathbf{s}^{(i)}$ is detected as $\mathbf{s}^{(j)}$, where $i \neq j$, with maximum likelihood detection over high SNR region is upper bounded by

$$P(\mathbf{s}^{(i)} \to \mathbf{s}^{(j)}) \leq \frac{1}{\|\mathbf{s}^{(i)} - \mathbf{s}^{(j)}\|^2} \left(\frac{P_{\text{tot}}}{4M_t N_0}\right)^{-M_r}. \tag{10.42}$$

As such, a diversity order of M_r is achieved. A brute force implementation of maximum likelihood detection requires the exhaustive search over all possible data symbol vectors $\mathbf{s}$. As such, the decoding complexity is exponential in the number of transmit antennas M_t. The detection complexity can be reduced with suboptimal linear detection schemes, such as zeroforcing (ZF) and minimium mean square error (MMSE) receivers, at the cost of a certain performance loss.

The basic idea of suboptimal linear detection is to linearly process the received symbol vector $\mathbf{y}$ such that each data symbol can be independently detected. ZF receiver completely removes mutual interference between data symbols during detection. In particular, the receiver first multiplies the received symbol vector with the pseudoinverse of the channel matrix $\mathbf{H}$ as

$$\mathbf{z} = \mathbf{H}^{\dagger}\mathbf{y} = (\mathbf{H}^H \mathbf{H})^{-1}\mathbf{H}^H \mathbf{y}, \tag{10.43}$$

which after substitution and simplification becomes

$$\mathbf{z} = \mathbf{s} + \mathbf{H}^{\dagger}\mathbf{n} = \mathbf{s} + \tilde{\mathbf{n}}. \tag{10.44}$$

Clearly, when $M_r \geq M_t$, ZF receiver decouples matrix channels into M_t parallel scalar channels with additive noise. Meanwhile, the noise on different scalar channels are correlated with enhanced noise power. It has been shown that the received SNR on the ith subchannel follows a Chi-squared distribution with $2(M_r - M_t + 1)$ degrees of freedom, with probability density function (PDF) given by

$$p(\gamma) = \frac{M_t}{(M_r - M_t)! P_{\text{tot}}/N_0} e^{-\frac{M_t \gamma}{P_{\text{tot}}/N_0}} \left(\frac{M_t \gamma}{P_{\text{tot}}/N_0}\right)^{M_r - M_t}. \tag{10.45}$$

As such, ZF receiver can achieve a diversity order of $M_r - M_t + 1$. Furthermore, ZF receiver suffers noise enhancement, which results in poor performance over low SNR region.

MMSE receiver improves the performance over low SNR regions by properly balancing the mutual interference mitigation and noise enhancement. Specifically, MMSE receiver processes the received symbol vector with the target of minimizing the mean square error (MSE) between its output symbols and transmitted data symbols, i.e., $\mathbf{E}[\|\mathbf{Gy} - \mathbf{s}\|^2]$, where $\mathbf{G}$ is the processing matrix. It can be shown that $\mathbf{G}$ for MMSE receiver is given by

$$\mathbf{G} = \arg \min_{\mathbf{G}} \mathbf{E}[\|\mathbf{Gy} - \mathbf{s}\|^2] = \left(\mathbf{H}^H \mathbf{H} + \frac{M_t}{P_{\text{tot}}/N_0} \mathbf{I}_{M_r}\right)^{-1} \mathbf{H}^H. \tag{10.46}$$

Over high SNR region, MMSE receiver converges to ZF receiver and achieves the same diversity order of $M_r - M_t + 1$.

10.3.2.2 Coded transmission

Channel coding techniques can improve the reliability of MIMO multiplexing transmission. Uncoded MIMO multiplexing transmission schemes in previous subsection can only achieve a maximum diversity order of M_r. With coding, full diversity gain can be achieved by spreading the coded bits of a codeword over both spatial and temporal dimensions. It has been shown that the minimum codeword length should be greater than the number of bits carried by $M_t M_r$ modulated symbols. Figure 10.6(a) shows a MIMO multiplexing transmission structure that can achieve full diversity. In particular, after the information bits are processed by encoder/modulator, the resulting coded symbols $[x_1, x_2, \ldots, x_N]$ of each codeword are converted into M_t parallel symbol streams by a serial-to-parallel converter. Each of these symbol streams, of length N/M_t, which is greater or equal to M_r, is transmitted from a distinct antenna. The transmitted symbols from M_t antennas over N/M_t symbol periods are arranged as

$$
\begin{array}{c|ccccc}
 & \text{1st } T_s & \text{2nd } T_s & \cdots & M_r\text{th } T_s & \cdots & N/M_t\text{th } T_s \\
\hline
\text{1st antenna} & x_1 & x_{M_t+1} & \cdots & x_{(M_r-1)M_t+1} & \cdots & \cdot \\
\text{2nd antenna} & x_2 & x_{M_t+2} & \cdots & x_{(M_r-1)M_t+2} & \cdots & \cdot \\
\vdots & \vdots & \vdots & \ddots & \vdots & \ddots & \vdots \\
M_t\text{th antenna} & x_{M_t} & x_{2M_t} & \cdots & x_{M_r M_t} & \cdots & x_N
\end{array}
\tag{10.47}
$$

Although this transmitter structure is simple, the receiver needs to perform joint detection on the received symbols on M_r antennas over N/M_t symbol periods. The decoding complexity at the receiver grows exponentially with the codeword length N, which renders the resulting transmission scheme impractical.

Several low-complexity spatial multiplexing transmission schemes were developed at Bell Labs, typically referred as BLAST structures for MIMO channels. The most notable ones are vertical BLAST (V-BLAST) and diagonal BLAST (D-BLAST) structures. With V-BLAST, the information bit stream is first demultiplexed into M_t independent substreams. These substreams will be independently encoded and modulated before being transmitted on M_t transmit antennas in parallel, as illustrated in Figure 10.6(b). The coded symbols corresponding to one codeword at the ith encoder/modulator output are $[x_1^{(i)}, x_2^{(i)}, \ldots, x_L^{(i)}]$, $i = 1, 2, \ldots, M_t$. The transmission from different antennas are perfectly synchronized with coded symbols spanning over the same L symbol periods. The coded/modulated substreams are vertical stacked together in a layered fashion, hence the name vertical BLAST.

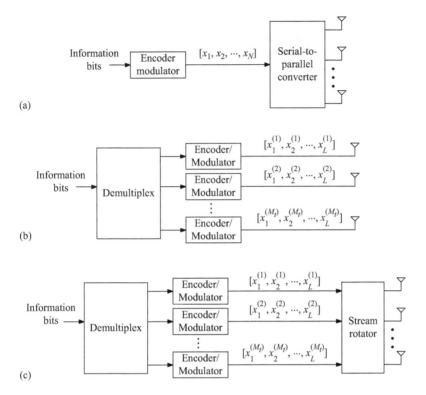

Figure 10.6 *Coded MIMO transmission: (a) coded MIMO transmission requiring joint detection, (b) coded MIMO transmission allowing sequential detection (V-BLAST), and (c) coded MIMO transmission allowing sequential detection (D-BLAST)*

The receiver for V-BLAST structure will perform detection on the received symbol vectors on M_r receive antennas over L symbol periods. Instead of applying maximum likelihood joint detection, the receiver will carry out sequential detection of each substream with successive interference cancelation (SIC), which has much lower complexity than joint detection. Specifically, the detection process works as follows: First, the M_t substreams are ordered according to their experienced signal-to-interference-plus-noise ratio (SINR) while treating other substreams as interference. Then, the substream with the highest SINR is first detected. After that, the detected substream is subtracted from the received signal. This process is repeated for the remaining $M_t - 1$ substreams. Such detection process is nonlinear due to the interference cancelation step. Typically, the number of receive antenna M_r should be greater than the number of substreams/transmit antennas for acceptable detection performance.

V-BLAST can achieve at most a diversity order of M_r, as each substream is transmitted from a single transmit antenna. Diagonal BLAST can realize the maximum diversity order of $M_t M_r$, with the introduction of a stream rotator. As illustrated in Figure 10.6(c), the demultiplexed information bit streams are first encoded/modulated in parallel, generating coded data symbols $[x_1^{(i)}, x_2^{(i)}, \ldots, x_L^{(i)}]$ for the ith stream. Then, the coded symbols are processed by a stream rotator, which cyclically changes the stream-antenna association. The rotation cycle is chosen such that the codewords of each stream are transmitted over all M_t antennas. As such, the length of symbol streams L should be at least M_t. The transmitted symbols from M_t antennas are arranged as

$$
\begin{array}{c|cccccccc}
 & 1\text{st } T_s & 2\text{nd } T_s & 3\text{rd } T_s & \cdots & M_t\text{th } T_s & M_t + 1\text{th } T_s & \cdots \\
\hline
\text{1st antenna} & x_1^{(1)} & x_1^{(2)} & x_1^{(3)} & \cdots & x_1^{(M_t)} & \cdot & \cdots \\
\text{2nd antenna} & & x_2^{(1)} & x_2^{(2)} & \cdots & x_2^{(M_t-1)} & x_2^{(M_t)} & \cdots \\
\text{3rd antenna} & & & x_3^{(1)} & \cdots & x_3^{(M_t-2)} & x_3^{(M_t-1)} & \cdots \\
\vdots & & & & \ddots & \vdots & \vdots & \ddots \\
M_t\text{th antenna} & & & & & x_{M_t}^{(1)} & x_{M_t}^{(2)} & \cdots
\end{array}
\tag{10.48}
$$

As a result, the coded/modulated substreams are distributed diagonally over the transmit antennas, hence the name.

The receiver for D-BLAST will sequentially detect and cancel the diagonal layers/substreams in a similar fashion as V-BLAST receiver. The initial lower triangular space-time resource block is not used to improve the reliability of top layer detection. The detection of the top layer will experience less interstream interference because of the unused triangular block, whereas the detection of first substream in V-BLAST will experience interference from all other $M_t - 1$ substreams. As such, D-BLAST enjoys better reliability than V-BLAST at the cost of certain space-time resource block wastage.

10.4 Diversity-multiplexing trade-off

The previous sections show that MIMO transmission can explore diversity gain and spatial multiplexing gain over MIMO wireless channels. The maximum possible diversity gain is $M_t M_r$, achievable with MRT/MRC or space-time diversity transmission combined with receiver MRC. The maximum spatial multiplexing gain is equal to the rank of the channel matrix $\mathbf{H}$, $R_{\mathbf{H}} \leq \min\{M_t, M_r\}$. When the channel matrix is full rank, then the multiplexing gain is limited by the minimum number of transmit and receive antennas. Full diversity gain and full multiplexing gain can rarely be achieved at the same time, unlike in certain special scenario. In general, there is a trade-off between diversity gain and multiplexing gain in real-world systems. In the following, we presented a simple characterization of this trade-off.

Let r denote the multiplexing gain achieved by a MIMO system, compared with a SISO system. For a M_t-by-M_r system, the value range of r is $[0, \min\{M_t, M_r\} - 1]$. For a given r value, the maximum diversity order d_0 that a MIMO transmission system can achieve is given by

$$
d_0(r) = (M_t - r)(M_r - r).
\tag{10.49}
$$

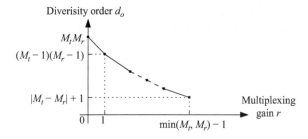

Figure 10.7 Diversity multiplexing trade-off

For example, the MIMO diversity transmission system with MRT/MRC achieves a diversity order of $M_t M_r$, while the multiplexing gain over single-input–single-output channel is 0. Meanwhile, the maximum diversity order of uncoded multiplexing transmission over a MIMO channel with $M_r \geq M_t$ can reach $M_r - M_t + 1$, as the multiplexing gain r is equal to $M_t - 1$. The diversity order versus multiplexing gain trade-off is illustrated in Figure 10.7.

10.5 Further readings

The pioneering work on MIMO systems can be found in [1,2]. Reference [3] presents further details about MIMO channel models. Reference [4, Chapter 10] summarized the capacity result on MIMO channels for various scenarios. Reference [5] is dedicated to space-time block coded transmission over MIMO channels. Further discussion on diversity multiplexing trade-off can be found in [6]. Reference [7] proves a comprehensive treatment on MIMO wireless transmission.

Problems

1. Consider a MIMO wireless channel with three transmit antennas and four receive antennas. The instantaneous channel gain matrix is given by

$$\mathbf{H} = \begin{bmatrix} 0.35 + j0.23 & 0.45 - j0.14 & 0.2 + j0.03 \\ 0.15 - j0.24 & 0.4 + j0.11 & 0.1 - j0.13 \\ 0.05 + j0.12 & 0.15 + j0.17 & 0.3 - j0.24 \\ 0.4 - j0.42 & 0.2 + j0.05 & 0.3 - j0.35 \end{bmatrix}. \tag{10.50}$$

Assuming the symbol energy over noise power ratio E_s/σ^2 is equal to 15 dB, determine the value of the instantaneous combined SNR at the receiver with TAS/MRC diversity.

2. Equal gain combining (EGC) can also apply to MIMO diversity transmission. Consider the 4-by-3 MIMO channel with instantaneous channel gain matrix given by

$$\mathbf{H} = \begin{bmatrix} 0.35 + j0.23 & 0.45 - j0.14 & 0.2 + j0.03 \\ 0.15 - j0.24 & 0.4 + j0.11 & 0.1 - j0.13 \\ 0.05 + j0.12 & 0.15 + j0.17 & 0.3 - j0.24 \\ 0.4 - j0.42 & 0.2 + j0.05 & 0.3 - j0.35 \end{bmatrix}. \tag{10.51}$$

Determine the combining weights at the receiver assuming that transmit antenna selection combined with EGC at the receiver is adopted.

3. Consider a MIMO wireless channel with four transmit antennas and three receive antennas. The instantaneous channel gain matrix is given by

$$\mathbf{H} = \begin{bmatrix} 0.35 & 0.45 & 0.2 & 0.4 \\ 0.15 & 0.4 & 0.15 & 0.2 \\ 0.05 & 0.15 & 0.3 & 0.3 \end{bmatrix}. \tag{10.52}$$

Assuming the transmit power over noise spectrum density ratio P_{tot}/N_0 is equal to 20 dB, determine the normalized instantaneous capacity of the channel C/B for (i) CSIT and (ii) CSI at the receiver only scenarios.

4. The instantaneous channel gain matrix of a MIMO channel with three transmit antennas and two receive antennas is given by

$$\mathbf{H} = \begin{bmatrix} 0.5 & -0.4 & 0.2 \\ -0.1 & 0.3 & 0.6 \end{bmatrix}. \tag{10.53}$$

(i) Determining the precoding matrix used at the transmitter to decompose the MIMO channel into two independent parallel channels.

(ii) Assuming the transmitter equally allocates the total transmit power $P_{tot} = 0.2$ watts to each channel, determine the capacity of this MIMO channel when the noise spectral density is $N_0 = 10^{-7}$ W/Hz and channel bandwidth 30 kHz.

(iii) Designing the optimal transmission strategy for maximal channel capacity, when the transmitter has full CSI.

5. Consider a MIMO channel with four transmit antennas and eight receive antennas. (i) If the transmission scheme achieved a multiplexing gain of 3, what is the maximum achievable diversity order, based on the diversity-multiplexing trade-off? (ii) If the desired diversity order is 5, what is the largest multiplexing gain we can expect?

Bibliography

[1] G. J. Foschini and M. Gans, "On limits of wireless communications in a fading environment using multiple antennas", *Wireless Pers. Commun.*, Vol. 6, pp. 311–335, March 1998.

[2] E. Telatar, "Capacity of multi-antenna Gaussian channels", *Euro. Trans. Telecommun.*, Vol. 10, pp. 585–596, November 1999.

[3] A. Paulraj, R. Nabar, and D. Gore, *Introduction to Space-Time Wireless Communications*, New York, NY: Cambridge University Press, 2003.

[4] A. Goldsmith, *Wireless Communications*, New York, NY: Cambridge University Press, 2005.

[5] E. G. Larsson and P. Stoica, *Space-Time Block Coding for Wireless Communications*, 2nd ed. New York, NY: Cambridge University Press, 2003.

[6] L. Zheng and D. N. Tse, "Diversity and multiplexing: A fundamental trade-off in multiple antenna channels", *IEEE Trans. Inform. Theory*, Vol. 49, pp. 2372–2388, October 2003.

[7] E. Biglieri, R. Calderbank, A. Constantinides, A. Goldsmith, A. Paulraj, and H. V. Poor, *MIMO Wireless Communications*, New York, NY: Cambridge University Press, 2007.

Chapter 11
Advanced wireless transmission

Multiple-antenna transmission and reception techniques can improve both the reliability and efficiency of digital wireless systems. Due to the size, cost, and complexity
constraints, it is in general challenging to implement multiple antennas at the mobile
terminals. Meanwhile, considering the antennas at different mobile terminals together,
we can create a multiple-antenna scenario, even if each terminal only has a single
antenna. In this chapter, we introduce several advanced wireless transmission technologies that explore in one way or another multiple antennas at different mobile
terminals to achieve diversity gain as well as multiplexing gain. In particular, we will
investigate multiuser diversity transmission, cooperative diversity transmission, and
multiuser multiple-input-multiple-output (MIMO) transmission technologies. Our
study of these advanced technologies will serve as the applications of the general
digital wireless transmission over fading channel framework that we established in
the previous chapters. Due to space limitation, these discussions will be by no means
comprehensive. Interested readers can refer to related literature for further details.

11.1 Multiuser diversity transmission

Let us first consider the fundamental scenario that a single-antenna base station is serving K single-antenna users, as shown in Figure 11.1. The wireless channels introduce

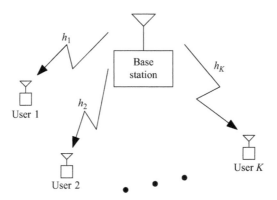

Figure 11.1   System model for multiuser diversity transmission

flat fading effects. The flat fading channel gain from the base station antenna to the kth user is denoted by h_k. With a given time/frequency channel resource, the conventional approach is to serve these K users in a round robin fashion, ensuring perfect fairness. Meanwhile, considering the K user antennas together, we have multiple antenna at the user side. To explore the potential diversity benefit, an alternative approach is to serve the user with the best instantaneous channel condition. The resulting transmission scheme is typically referred to as *multiuser diversity transmission* or *user scheduling.*

The received signal-to-noise ratio (SNR) of each user can serve as the quality indicator for the user channel. The instantaneous received SNR at user k is given by

$$\gamma_k = \frac{|h_k|^2 E_s}{N_0}, \quad k = 1, 2, \ldots, K, \tag{11.1}$$

where E_s is the symbol energy of the transmitted signal and N_0 is the power spectrum density of the additive noise at the receiver. Typically, γ_ks are independent random variables as users are geographically distributed. For the Rayleigh fading environment, γ_k will be an exponential random variable with probability density function (PDF) given by

$$p_{\gamma_k}(\gamma) = \frac{1}{\overline{\gamma}_k} \exp\left(-\frac{\gamma}{\overline{\gamma}_k}\right), \quad \gamma > 0, \tag{11.2}$$

where $\overline{\gamma}_k$ is the average received SNR of user k, whose value depends upon the corresponding path loss and shadowing effects.

11.1.1 Scheduling strategies

A basic user scheduling strategy is to schedule the user with the largest instantaneous received SNR for service. Mathematically, the index of the scheduled user is given by

$$k^* = \arg\max_k \{\gamma_k\}. \tag{11.3}$$

As such, the scheduled user's SNR will be the largest one among K-independent user SNRs. If the user channels experience identical Rayleigh fading, i.e., $\overline{\gamma}_1 = \overline{\gamma}_2 = \cdots = \overline{\gamma}_K = \overline{\gamma}$, then the PDF of the scheduled user's SNR can be shown to be given by

$$p_{\gamma_{k^*}}(\gamma) = \frac{K}{\overline{\gamma}} \left[1 - \exp\left(-\frac{\gamma}{\overline{\gamma}}\right)\right]^{K-1} \exp\left(-\frac{\gamma}{\overline{\gamma}}\right). \tag{11.4}$$

The system achieves the same diversity benefit of a K-branch selection combining scheme. In practice, different users will experience different path loss/shadowing effects as well as distinct fading severity. Such basic user-scheduling strategy will be highly unfair as those users located closer to the base station will be scheduled more frequently than those further away from the base station.

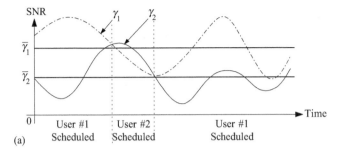

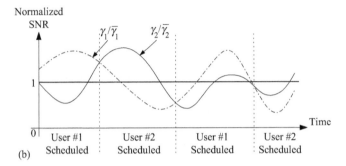

Figure 11.2 Operation of SNR-based user scheduling strategies: (a) absolute SNR-based user scheduling and (b) normalized SNR-based user scheduling

An alternative user scheduling strategy is to schedule the user with the largest normalized instantaneous SNR. In particular, the instantaneous user SNR γ_k is normalized by its average $\overline{\gamma}_k$ before being compared. Mathematically, the index of the scheduled user with normalized SNR-based scheduling is given by

$$k^* = \arg\max_k \left\{ \frac{\gamma_k}{\overline{\gamma}_k} \right\}. \tag{11.5}$$

The unfairness due to different path loss/shadowing effect is mitigated. It can be shown that such normalized SNR-based user scheduling can achieve long-term fairness among users, at the cost of slightly poor performance compared with absolute SNR-based scheduling. Figure 11.2 illustrates the operation of the absolution SNR and normalized SNR-based user scheduling strategies. User 1 enjoys better channel condition on average and as such will be scheduled much more often than user 2 with absolute SNR-based scheduling. The normalized SNR-based user scheduling leads to better fairness between users.

Example: Multiuser scheduling

A base station is serving three users over a wireless channel with the same transmit power. The average received SNRs at the users, considering path loss, shadowing, and noise effects, are equal to 9, 15, and 8 dB. At a particular time instance, the received SNRs are 11, 14, and 9 dB. Determine the received SNR of the scheduled user at this time instance with

1. absolute SNR-based and
2. normalized SNR-based user scheduling strategies.

Solutions:

1. With absolute SNR-based user scheduling, the base station will compare the instantaneous SNR of different users. In this case, the second user has the highest instantaneous SNR. As such, the second user will be scheduled, leading to the scheduled user SNR of 14 dB.
2. With normalized SNR-based user scheduling, the base station will compare the normalized instantaneous SNR of different users, which are 2 , −1, and 1 dB. The first user has the highest normalized SNR and will be scheduled. The scheduled user's SNR is 11 dB.

The fairness among users can also be improved if we take into consideration of the historical throughput of different users. The resulting *proportional fair scheduling* strategy selects the user with the largest normalized instantaneous throughput for transmission. Specifically, given the instantaneous receiver SNR γ_k over the *i*th scheduling interval, the instantaneous throughput of user k is calculated using Shannon's formula as

$$C_k(i) = \log_2(1 + \gamma_k). \tag{11.6}$$

The instantaneous throughput is then normalized by the historical throughput of user k up to the previous time slot, denoted by $R_k(i)$. Mathematically, the index of the scheduled user over the *i*th slot is given by

$$k^* = \arg\max_k \left\{ \frac{C_k(i)}{R_k(i)} \right\}. \tag{11.7}$$

The system will regularly update the historical throughput $R_k(i)$ as

$$R_k(j + 1) = R_k(j), \qquad k \neq k^*, \tag{11.8}$$

$$R_k(j + 1) = R_k(j) + C_k(j), \quad k = k^*.$$

Again, it can be theoretically shown that proportional fair scheduling can achieve long-term fairness.

11.1.2   Implementation issues

Multiuser diversity transmission can explore diversity benefit over multiple user channels. Compared to traditional antenna diversity system, multiuser diversity has several unique advantages. In particular, the receiver has low complexity, as each user only needs a single antenna. The channel experienced by multiple users is naturally independent as the users are distributed over the service area of the base station. Conventional diversity systems try to eliminate deep fade by transmitting or receiving copies of the same signal, whereas multiuser diversity exploits the good condition of user channels.

The diversity gain from user scheduling comes at the cost of a certain fairness issue. Multiuser diversity system can only achieve long-term fairness through SNR normalization or using historical throughput information. We need to properly balance a trade-off between diversity gain and fairness among users. Another challenge facing multiuser diversity transmission system is the requirement of a centralized knowledge of user channel condition. In particular, the system needs to collect the quality information of all user channels, usually at the base station in order to determine the scheduled user. For uplink transmission, the base station may obtain the channel quality information based on the pilots sent from the users. These processes will introduce a certain delay and increase system complexity and overhead. For downlink transmission, the base station has to rely on user feedback to collect user channel quality information, if the channel does not exhibit reciprocity property as in time division duplexing (TDD) systems. Specifically, the base station sends a pilot and each user estimates and feeds back their channel quality information. The feedback load involved will be K real numbers over each scheduling period. Note that the scheduling period should be less than the channel coherence time T_c. As such, excess feedback overhead will undermine the performance gain with user scheduling.

An effective approach to reduce the feedback load for multiuser diversity downlink transmission is to allow only those users with good enough channel quality to feedback. The rationale behind this approach is that (i) the scheduler only needs to know the user with the best quality, which can be determined if as least one user feeds back and (ii) the users with poor channel condition have little chance to be scheduled, and as such, they can hold their feedback when their channel condition improves. During each scheduling period, users will compare their channel quality indicator, e.g. $\gamma_k/\overline{\gamma}_k$ for normalized SNR-based scheduling, with a predetermined threshold, denoted by η_{th}, and feedback their channel quality only if $\gamma_k/\overline{\gamma}_k > \eta_{\text{th}}$. The base station will schedule the user that feeds back the largest normalized SNR. It may happen in the worst case that no user feedbacks, in which case, the base station may schedule a random user. In summary, the scheduled user's received SNR with this strategy is equal to γ_{k^*}, where $k^* = \arg\max_k \left\{ \frac{\gamma_k}{\overline{\gamma}_k} \right\}$, if at least one user feedback and γ_k, where k is the index of a random user, if no user feedback.

11.2 Cooperative relay transmission

Relay transmission has been widely used in satellite communications and terrestrial microwave communications. The main objective was to extend the transmission distance. Relay transmission was introduced to cellular wireless systems to improve the coverage of cell-edge users. It serves as an efficient solution to improve the

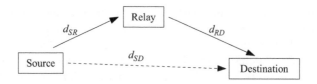

Figure 11.3   Relay transmission system

quality of service for mobile users who experience severe path loss/shadowing effect. Furthermore, cooperative relay transmission can achieve a certain diversity gain by exploring multiple antennas at different users.

The basic structure of a relay transmission system is shown in Figure 11.3. Specifically, the source node is sending information to the destination node with the help of the relay node. The participation of the relay node is especially valuable when the direct source-to-destination link has poor quality. Given the broadcasting nature of most wireless transmission, relay node can overhear the transmitted signal from the source and forward a copy to the destination after proper processing. Amplify-and-forward (AF) and decode-and-forward (DF) are two of the most common relaying strategies.

11.2.1   Amplify-and-forward relaying

Let us assume that the source node transmits signal $s(t)$ with symbol energy E_s. The wireless channels introduce flat fading effect. The received signal at the relay node is given by

$$r_1(t) = h_1 s(t) + n_1(t), \tag{11.9}$$

where h_1 is the complex channel gain of the source–relay link and $n_1(t)$ is the additive white Gaussian noise (AWGN) at relay node with power spectrum density $N_0/2$. With AF relaying, the relay node will amplify the received signal with a gain G and then forward it to the destination. The received signal at the destination node is given by

$$r_2(t) = h_2 G(h_1 s(t) + n_1(t)) + n_2(t) = h_2 G h_1 \cdot s(t) + h_2 G \cdot n_1(t) + n_2(t), \tag{11.10}$$

where h_2 is the complex channel gain of the relay–destination link, and $n_2(t)$ is the AWGN at the destination node. The received SNR at the destination node can be determined as

$$\gamma^{AF} = \frac{|h_1|^2 |h_2|^2 G^2 E_s}{|h_2|^2 G^2 N_0 + N_0}. \tag{11.11}$$

Typically, the power amplifying gain G at the relay is set to satisfy the output power constraint of the relay node. Specifically, the output power of the relay node, given by $G^2 |h_1|^2 E_s + G^2 N_0$, should be no more than a certain maximum value. Let us assume that G is chosen such that the output symbol energy at the relay is also equal to E_s. It can be shown that

$$G^2 = \frac{E_s}{|h_1|^2 E_s + N_0}. \tag{11.12}$$

The resulting received SNR at the destination becomes

$$\gamma^{AF} = \frac{|h_1|^2 E_s/N_0 \cdot |h_2|^2 E_s/N_0}{|h_1|^2 E_s/N_0 + |h_2|^2 E_s/N_0 + 1} = \frac{\gamma_1 \gamma_2}{\gamma_1 + \gamma_2 + 1}, \tag{11.13}$$

where γ_1 and γ_2 are the received SNRs of the first and second hops, respectively. The performance of relay transmission with AF relaying can be analyzed with the statistics of γ^{AF}. Over high SNR region, where $\gamma_1 + \gamma_2 \gg 1$, the received SNR with AF relaying can be approximated by

$$\gamma^{AF} \approx \frac{\gamma_1 \gamma_2}{\gamma_1 + \gamma_2}, \tag{11.14}$$

which can lead to simpler statistical characterization of γ^{AF} for performance analysis.

11.2.2 Decode-and-forward relaying

With DF relaying, the relay node first demodulates and decodes the source signal. Then, the relay will forward a re-encoded copy of the source signal to the destination. The relay transmission link can be viewed as a cascade connection of two independent wireless links. As such, the performance of the overall relay transmission with DF relaying is limited by the weaker link. The quality indicator of the relay link becomes the minimum of two hop SNRs, i.e., $\min\{\gamma_1, \gamma_2\}$.

To further illustrate, let us consider the outage probability of a DF relaying transmission link. Outage of a DF relay link occurs when either the source–relay link or relay–destination link has very low received SNR, i.e., below the outage threshold γ_{th}. As such, the outage probability is calculated as

$$P_{out} = 1 - \Pr[\text{no hop in outage}] = 1 - \Pr[\gamma_1 > \gamma_{th}, \gamma_2 > \gamma_{th}], \tag{11.15}$$

which can be shown to be equal to

$$P_{out} = \Pr[\min\{\gamma_1, \gamma_2\} < \gamma_{th}]. \tag{11.16}$$

For Rayleigh fading environment, the PDF of $\min\{\gamma_1, \gamma_2\}$, given by

$$p_{\gamma_{eq}^{DF}}(\gamma) = \frac{1}{\overline{\gamma}_{min}} e^{-\gamma/\overline{\gamma}_{min}}, \tag{11.17}$$

where $\overline{\gamma}_{min} = \frac{\overline{\gamma}_1 \overline{\gamma}_2}{\overline{\gamma}_1 + \overline{\gamma}_2}$, can be readily applied to evaluate the outage probability.

Compared to AF relaying, DF relaying introduces additional complexity for relay decoding/encoding process. Such additional complexity helps avoid the forwarding of noise signal with AF relaying. Meanwhile, the relay node with DF relaying may forward undetected errors to the destination and cause error propagation, especially when the source–relay link quality is poor.

11.2.3   Cooperative diversity

Transmission through the relay node leads to an alternative signaling path. If the destination node can receive information over both the source–destination link and the relay link, then the resulting *cooperative relay* transmission system can enjoy a certain diversity benefit. Typically, cooperative relay transmission is carried out over two consecutive time slots. During the first time slot, the source node broadcasts information signal to relay and destination. Over the second time slot, the relay node transmits to the destination node. Let us assume that the received SNR of direct source–destination link is γ_0. With AF relaying, the received SNR through relay link is $\frac{\gamma_1\gamma_2}{\gamma_1+\gamma_2+1}$. If the destination applies maximum ratio combining (MRC) on the received signal copies over two time slots, the overall received SNR is given by

$$\gamma_c = \gamma_0 + \frac{\gamma_1\gamma_2}{\gamma_1 + \gamma_2 + 1}. \tag{11.18}$$

The diversity gain of dual branch MRC receiver is realized. Meanwhile, the cooperative diversity gain comes at the cost of a certain spectral efficiency loss. Note that with direct transmission only, the source can transmit information over both time slots. With cooperative relaying, the source can only transmit for half of the duration. Such loss originates from the fact that most relay nodes operate in a half-duplex fashion.

   Incremental relaying can reduce such spectral efficiency loss while still enjoying certain diversity benefit. With incremental relaying transmission, the source node broadcasts information signal over the first time slot, while both relay and destination nodes receive the transmission. After the first time slot, the destination node will decide if it can decode the source transmission. If "yes," the destination will send an acknowledgment to the source and relay nodes. The relay node, upon receiving this acknowledgment, will discard the information it received, and the source node will proceed to transmit new information. If the destination cannot decode the source information, it will inform the relay to forward its received signal copy and the source node to hold its transmission of new information. The destination node will perform MRC on the received signal over both links. Assuming the destination node can successfully detect the information over direct transmission link if and only if the direct link SNR satisfies $\gamma_0 \geq \gamma_{th}$, the overall received SNR of incremental relaying is given by

$$\gamma_c^{IR} = \begin{cases} \gamma_0 + \frac{\gamma_1\gamma_2}{\gamma_1+\gamma_2+1}, & \gamma_0 < \gamma_{th}; \\ \gamma_0, & \gamma_0 \geq \gamma_{th}. \end{cases} \tag{11.19}$$

The system achieves a similar diversity benefit of the dual-branch switch and stay combining (SSC) scheme.

Example: Cooperative relay

Consider a three-node relay transmission scenario, where the source and relay transmit with the same output power. The relay adopts amplify and forward relaying. The wireless channel introduces flat fading, which leads to the instantaneous received SNR over source–destination link, source–relay link, and relay–destination link equal to 8, 15, and 12 dB, respectively. Determine the received SNR over the relay link. What is the overall combined SNR at the receiver when the system adopts cooperative relaying with MRC combining at the destination?

Solutions: The received SNR over the relay link with AF relaying can be determined as

$$\gamma^{AF} = \frac{\gamma_1 \gamma_2}{\gamma_1 + \gamma_2 + 1} = \frac{31.6 \times 15.8}{31.6 + 15.8 + 1} = 10.3 = 10.1 \text{ dB}. \tag{11.20}$$

The approximate SNR formula leads to $\frac{31.6 \times 15.8}{31.6 + 15.8} = 10.5$, which is slightly higher than the exact one.

The overall received SNR with cooperative relaying is equal to

$$\gamma_c = 6.3 + \frac{31.6 \times 15.8}{31.6 + 15.8 + 1} = 16.6 = 12.2 \text{ dB}. \tag{11.21}$$

11.3 Multiuser MIMO transmission

When the base station is equipped with multiple antennas, we arrive at a typical scenario of a multiple-antenna base station serving K single-antenna users, as illustrated in Figure 11.4. Let us assume that the base station has M antennas, $M < K$. The user channel is then characterized by a channel vector of length M. Specifically, the channel vector of the kth user is $\mathbf{h}_k = \{h_{k1}, h_{k2}, \ldots, h_{kM}\}^T$, where h_{ki} denotes the complex channel gain from the ith base station antenna to user k. If a single user is selected to be served by the base station, then the system can explore the diversity gain from multiple base station antennas and possibly multiuser diversity gains through user scheduling. On the other hand, if we consider the multiple user antennas jointly, we arrive at a special MIMO system, with multiple antennas at the base station and multiple antennas at the user side. The resulting system is usually referred as

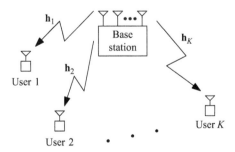

Figure 11.4   System model for multiuser MIMO transmission

multiuser MIMO systems. The main difference from the conventional point-to-point MIMO system is that the antennas at the user side are distributed. Various transmission schemes have been proposed for achieving spatial multiplexing gain over such multiuser MIMO channels. In general, the downlink transmission is more challenging as the users typically cannot collaborate during reception. For uplink transmission, the base station can perform multiuser joint detection based on the signal received on all antennas. We present several sample transmission schemes for multiuser MIMO downlink transmission in this section.

11.3.1   Linear zero forcing beamforming

Let x_i, $i = 1, 2, \ldots, M$, denote the transmitted symbol from the ith base station antenna over a symbol period. The corresponding received symbol at user k is given by

$$y_k = h_{k1}x_1 + h_{k2}x_2 + \cdots + h_{kM}x_M + n_k, \tag{11.22}$$

where n_k is the sample of the AWGN. If x_1 is the data symbol that user k wants to detect, then $h_{k2}x_2 + \cdots + h_{kM}x_M$ will act as multiuser interference and degrade the detection performance. To transmit data symbols to multiple user effectively, the base station may apply certain precoding operations to the data symbols, typically referred as beamforming transmission. We concentrate linear beamforming scheme here.

With linear beamforming, the transmitted symbols from different antennas are linear combinations of data symbols to be transmitted. The structure of a linear beamforming transmitter is shown in Figure 11.5. Specifically, the transmitted symbol from ith antenna, x_i, is given by

$$x_i = b_{1i}s_1 + b_{2i}s_2 + \cdots + b_{Mi}s_M, \quad i = 1, 2, \ldots, M, \tag{11.23}$$

where $s_j, j = 1, 2, \ldots, M$ are the data symbols and b_{ji}s are the beamforming weights. Note that these weights should be properly normalized to satisfy the output power constraint of the ith transmit antenna. If we group x_is from different antenna together in a vector form, we have

$$\mathbf{x} = \begin{bmatrix} x_1 \\ x_2 \\ \vdots \\ x_M \end{bmatrix} = \begin{bmatrix} b_{11} \\ b_{12} \\ \vdots \\ b_{1M} \end{bmatrix} \cdot s_1 + \begin{bmatrix} b_{21} \\ b_{22} \\ \vdots \\ b_{2M} \end{bmatrix} \cdot s_2 + \cdots + \begin{bmatrix} b_{M1} \\ b_{M2} \\ \vdots \\ b_{MM} \end{bmatrix} \cdot s_M \tag{11.24}$$

$$= \mathbf{b}_1 \cdot s_1 + \mathbf{b}_2 \cdot s_2 + \cdots + \mathbf{b}_M \cdot s_M.$$

It follows that the received symbol at user k is given by

$$y_k = \mathbf{h}_k^T \cdot \mathbf{x} + n_k = \mathbf{h}_k^T \mathbf{b}_1 \cdot s_1 + \mathbf{h}_k^T \mathbf{b}_2 \cdot s_2 + \cdots + \mathbf{h}_k^T \mathbf{b}_M \cdot s_M + n_k. \tag{11.25}$$

Assuming without loss of generality that user k is interested in data symbol s_k, we can rewrite the received symbol as

$$y_k = (\mathbf{h}_k^T \mathbf{b}_k)s_k + \sum_{m=1, m \neq k}^{M} (\mathbf{h}_k^T \mathbf{b}_m)s_m + n_k. \tag{11.26}$$

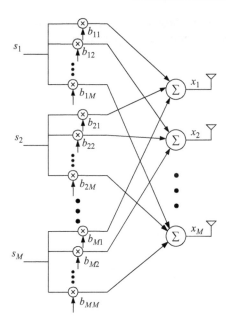

Figure 11.5 Structure of a linear beamforming transmitter

The second term on the right hand side $\sum_{m=1,m\neq k}^{M} (\mathbf{h}_k^T \mathbf{b}_m) s_m$ acts as multiuser interference during the detection of s_k based on y_k.

The multiuser interference can be completely eliminated with the so-called zero-forcing (ZF) beamforming strategy. In particular, if we design $\mathbf{b}_m$s such that $\mathbf{h}_k^T \mathbf{b}_m = 0$ for $m = 1, 2, \ldots, M, m \neq k$, then the received symbol at user k becomes

$$y_k = (\mathbf{h}_k^T \mathbf{b}_k) s_k + n_k. \tag{11.27}$$

As such, s_k can be detected with zero multiuser interference, hence the name zero-forcing. To achieve zero multiuser interference reception at all M users, we need

$$\begin{bmatrix} \mathbf{h}_1^T \\ \mathbf{h}_2^T \\ \vdots \\ \mathbf{h}_M^T \end{bmatrix} \cdot \begin{bmatrix} \mathbf{b}_1 & \mathbf{b}_2 & \cdots & \mathbf{b}_M \end{bmatrix} = \begin{bmatrix} \mathbf{h}_1^T \mathbf{b}_1 & 0 & \cdots & 0 \\ 0 & \mathbf{h}_2^T \mathbf{b}_2 & \cdots & 0 \\ \vdots & \vdots & \ddots & \vdots \\ 0 & 0 & \cdots & \mathbf{h}_M^T \mathbf{b}_M \end{bmatrix}. \tag{11.28}$$

The values of $\mathbf{h}_k^T \mathbf{b}_k$s may be chosen according to certain power allocation strategy. As such, the beamforming vectors $\mathbf{b}_m$s are related to the columns of the pseudoinverse of the channel gain matrix $\mathbf{H}$, given by

$$
\mathbf{H} = \begin{bmatrix} \mathbf{h}_1^T \\ \mathbf{h}_2^T \\ \vdots \\ \mathbf{h}_M^T \end{bmatrix} = \begin{bmatrix} h_{11} & h_{12} & \cdots & h_{1M} \\ h_{21} & h_{22} & \cdots & h_{2M} \\ \vdots & \vdots & \ddots & \vdots \\ h_{M1} & h_{M2} & \cdots & h_{MM} \end{bmatrix}. \tag{11.29}
$$

For perfect multiuser interference removal, the base station needs to have complete and accurate knowledge of the gain matrix of multiuser MIMO channel. In practice, making downlink channel matrix $\mathbf{H}$ available at the base station will lead to considerable overhead. For frequency division duplexing (FDD) systems, the users need to feed back their estimated channel vectors, which involves a feedback load of M complex number per user. For TDD systems, each user needs to send pilot symbols over uplink transmission to help the base station estimate downlink channel gains.

Example: ZF beamforming over multiuser MIMO channel

Consider a multiuser MIMO transmission scenario where the base station uses four transmit antennas to serve four mobile users. The instantaneous channel gain matrix is given by

$$
\mathbf{H} = \begin{bmatrix} \mathbf{h}_1^T \\ \mathbf{h}_2^T \\ \mathbf{h}_3^T \\ \mathbf{h}_4^T \end{bmatrix} = \begin{bmatrix} 0.35+j0.23 & 0.45-j0.14 & 0.2+j0.03 & 0.4-j0.42 \\ 0.5-j0.04 & 0.14+j0.3 & 0.13-j0.2 & 0.02+j0.5 \\ 0.15-j0.24 & 0.4+j0.11 & 0.1-j0.13 & 0.2+j0.05 \\ 0.05+j0.12 & 0.15+j0.17 & 0.3-j0.24 & 0.3-j0.35 \end{bmatrix}. \tag{11.30}
$$

Assuming that ZF beamforming transmission scheme is adopted at the base station with equal power allocation to different users, determine the beamforming vectors for each user.

Solutions: With ZF beamforming and equal power allocation, the beamforming vectors should satisfy

$$
\mathbf{H} \cdot \begin{bmatrix} \mathbf{b}_1 & \mathbf{b}_2 & \mathbf{b}_3 & \mathbf{b}_4 \end{bmatrix} = \begin{bmatrix} c & 0 & 0 & 0 \\ 0 & c & 0 & 0 \\ 0 & 0 & c & 0 \\ 0 & 0 & 0 & c \end{bmatrix} = c\mathbf{I}_4, \tag{11.31}
$$

where c is a normalizing constant to satisfy the total transmit power constraint. Therefore, the vectors $\mathbf{b}_i$ should be the columns of $c\mathbf{H}^{-1}$, which are determined as

$$
\begin{bmatrix} \mathbf{b}_1 & \mathbf{b}_2 & \mathbf{b}_3 & \mathbf{b}_4 \end{bmatrix} = \tag{11.32}
$$

$$
c \cdot \begin{bmatrix}
0.99 - j0.66 & 0.73 - j0.87 & -1.3 + j1.5 & -0.49 - j0.25 \\
0.37 + j0.68 & -0.49 + j1.15 & 2.74 - j0.86 & -0.86 - j0.52 \\
-0.28 - j0.82 & 0.13 + j1.73 & 1.97 - j1.41 & 0.72 + j1.82 \\
0.42 + j0.25 & 0.56 - j1.41 & -1.85 - j0.02 & 0.66 + j0.51
\end{bmatrix}.
$$

11.3.2   Spatial multiplexing with random beamforming

Random beamforming can achieve a certain spatial multiplexing gain over multiuser MIMO channel with reduced feedback. With random beamforming, the base station with M antennas can serve up to M users using randomly orthonormal beamforming vectors. Let $\{\mathbf{u}_1, \mathbf{u}_2, \ldots, \mathbf{u}_M\}$ denote the set of random beamforming vectors, that satisfies $\|\mathbf{u}_m\|^2 = 1$, $m = 1, 2, \ldots, M$, and $\mathbf{u}_i^T \mathbf{u}_j = 0$, for $i \neq j$. The transmitted symbol vector is generated as

$$
\mathbf{x} = \mathbf{u}_1 \cdot s_1 + \mathbf{u}_2 \cdot s_2 + \cdots + \mathbf{u}_M \cdot s_M, \tag{11.33}
$$

where s_ms are data symbols. It follows that the received symbol at user k is given by

$$
y_k = \mathbf{h}_k^T \mathbf{u}_1 \cdot s_1 + \mathbf{h}_k^T \mathbf{u}_2 \cdot s_2 + \cdots + \mathbf{h}_k^T \mathbf{u}_M \cdot s_M + n_k. \tag{11.34}
$$

Assuming without loss of generality that user k is interested in data symbol s_j, y_k can be rewritten as

$$
y_k = (\mathbf{h}_k^T \mathbf{u}_j)s_j + \sum_{m=1,m\neq j}^{M} (\mathbf{h}_k^T \mathbf{u}_m)s_m + n_k, \tag{11.35}
$$

where $\sum_{m=1,m\neq j}^{M} (\mathbf{h}_k^T \mathbf{u}_m)s_m$ acts as multiuser interference, negatively affecting the detection of s_j. The detection performance will depend on the resulting signal-to-interference-plus-noise ratio (SINR), given by

$$
\gamma_{k,j} = \frac{|\mathbf{h}_k^T \mathbf{u}_j|^2}{\sum_{m=1,m\neq j}^{M} |\mathbf{h}_k^T \mathbf{u}_m|^2 + N_0/E_s}. \tag{11.36}
$$

Random beamforming transmission mitigates the effect of multiuser interference through user scheduling. In particular, each user can calculate its experienced SINR on different beamforming directions with the knowledge of the beamforming vectors and the estimated channel vector. The users will feed back their received SINR on M beams, based on which the base station assigns a beamforming direction to the user that enjoys the largest SINR on it. Mathematically speaking, user i will be assigned to beamforming direction $\mathbf{u}_j$ if

$$
\gamma_{i,j} \geq \gamma_{k,j}, \quad k = 1, 2, \ldots, K. \tag{11.37}
$$

As such, the received SINR of scheduled user on the jth beam is given by

$$\gamma_{j^*} = \max_k \{\gamma_{k,j}\}. \tag{11.38}$$

It has been theoretically shown that random beamforming with such user-scheduling strategy achieves similar sum rate performance as ZF beamforming scheme, when K is sufficiently large. Meanwhile, with random beamforming, each user only needs to feed back M real numbers.

The feedback load of random beamforming can be further reduced by noting that if a user's received SINR on beamforming direction $\mathbf{u}_j$ is not the largest for this user, then the chance that this user is assigned with $\mathbf{u}_j$ will be small. As such, it is usually sufficient for each user to feed back its largest SINR value and the index of the beam that leads to this SINR value. Specifically, each user will rank the SINR values on M beamforming directions and identify the best beamforming direction for its reception. Then, the user will feed back the best beam SINR, which for user i is given by

$$\gamma_i^* = \max_j \{\gamma_{i,j}\}, \tag{11.39}$$

and the corresponding beam index, given by

$$j^* = \arg\max_j \{\gamma_{i,j}\}. \tag{11.40}$$

The resulting feedback load will be one real number and one integer per user. The base station will rank all the feedback SINRs for each beamforming direction and assign the beams to the user with the largest feedback SINR on them. When the number of users is large enough and the probability that a beam is not the best beam for any user is near zero, such best beam SINR and index feedback strategy have the same performance as the strategy where each user feeds back M SINR values. There are several other feedback and user-scheduling strategies for random beamforming, which leads to different trade-off between feedback load and system performance.

11.4 Further readings

Early works on multiuser diversity transmission are presented in [1–3]. The performance analysis of multiuser diversity transmission systems is investigated in [4,5]. Reference [6, Chapter 6] presents further details on multiuser parallel scheduling transmission. Early works on cooperative relay transmission include [7–9]. More comprehensive treatment on cooperative communication can be found in [10,11]. Early works on multiuser MIMO channels include [12–14]. The optimality of ZF beamforming was established in [15–17]. Random beamforming was first considered in [18–20]. Reference [6, Chapter 7] presents further details on the multiuser MIMO transmission schemes.

Problems

1. Consider a multiuser diversity transmission system where the base station is serving four users using a single wireless channel. The average received SNRs at the users are equal to 9, 15, 12, and 8 dB. At a particular time instance, the user received SNR are 11, 14, 10, and 9 dB. A user will feed back its normalized received SNR if it is greater than 1.2 dB. Determine the received SNR of the scheduled user. What if at another time instant, the instantaneous received SNR at the users become 10, 15, 14, and 11.5 dB.

2. Consider a three-node relay transmission scenario, where the source and the relay transmit with the same output power. The system adopts incremental relaying with amplify and forward relaying strategy. The instantaneous received SNR over source–destination link, source–relay link, and relay–destination link are equal to 8, 15, and 12 dB, respectively. Determine the overall received SNR if the SNR threshold at the destination node is 9 dB. What if at another time instant, the instantaneous received SNR over three links become 10, 14, and 11.5 dB?

3. Consider a multiuser MIMO transmission scenario where the base station has two antennas, and the four mobile users have single antennas. The base station adopts random beamforming with beamforming vectors $\mathbf{u}_1 = [0.96, 0.28]^T$ and $\mathbf{u}_2 = [-0.280.96]^T$. The instantaneous channel gain matrix is given by

$$\mathbf{H} = \begin{bmatrix} \mathbf{h}_1^T \\ \mathbf{h}_2^T \\ \mathbf{h}_3^T \\ \mathbf{h}_4^T \end{bmatrix} = \begin{bmatrix} 0.35 + j0.23 & 0.45 - j0.14 \\ 0.5 - j0.04 & 0.14 + j0.3 \\ 0.15 - j0.24 & 0.4 + j0.11 \\ 0.05 + j0.12 & 0.15 + j0.17 \end{bmatrix}. \tag{11.41}$$

Assuming each user will feed back its received SINR on both beamforming directions, determine the scheduled users for each beamforming direction by assuming a certain value for E_b/N_0. Will the results be affected when E_b/N_0 value changes. What if the users only feed back the SINR of their better beam and its index?

Bibliography

[1] R. Knopp and P. Humblet, "Information capacity and power control in single-cell multiuser communications", in *Proc. IEEE Int. Conf. Commun. (ICC'95)*, Seattle, WA, vol. 1, pp. 331–335, June 1995.

[2] D. N. C. Tse, "Optimal power allocation over parallel Gaussian channels", in *Proc. Int. Symp. Inform. Theory (ISIT'97)*, Ulm, Germany, p. 27, June 1997.

[3] P. Viswanath, D. Tse, and R. Laroia, "Opportunistic beamforming using dumb antennas", *IEEE Trans. Inf. Theory*, vol. 48, pp. 1277–1294, June 2002.

[4] D. Gesbert and M.-S. Alouini, "How much feedback is multi-user diversity really worth?", in *Proc. of IEEE Int. Conf. on Commun. (ICC'04)*, Paris, France, June 2004, pp. 234–238.

[5]   L. Yang and M.-S. Alouini, "Performance analysis of multiuser selection diversity", *IEEE Trans. Veh. Technol.*, vol. 55, pp. 1003–1018, May 2006.

[6]   H.-C. Yang and M.-S. Alouini, *Order Statistics in Wireless Communications*, New York, NY: Cambridge University Press, 2011.

[7]   A. Sendonaris, E. Erkip, and B. Aazhang, "User cooperation diversity part I: system description", *IEEE Trans. Commun.*, vol. 51, pp. 1927–1938, 2003.

[8]   J. N. Laneman, D. N. C. Tse, and G. W. Wornell, "Cooperative diversity in wireless networks: efficient protocols and outage behavior", *IEEE Trans. Inf. Theory*, vol. 50, pp. 3062–3080, 2004.

[9]   A. Nosratinia, T. Hunter, and A. Hedayat, "Cooperative communication in wireless networks", *IEEE Commun. Mag.*, vol. 42, pp. 68–73, 2004.

[10] M. Dohler and Y. Li, *Cooperative Communications*, New York, NY: John Wiley & Sons, 2010.

[11] P. Y.-W. Hong, W.-J. Huang, and J. C.-C. Kuo, *Cooperative Communication and Networking*, New York, NY: Springer, 2010.

[12] G. Caire and S. Shamai, "On the achievable throughput of a multiantenna Gaussian broadcast channel", *IEEE Trans. Inf. Theory*, vol. IT-49, no. 7, pp. 1691–1706, July 2003.

[13] P. Viswanath and D. Tse, "Sum capacity of the vector Gaussian broadcast channel and uplink-downlink duality", *IEEE Trans. Inf. Theory*, vol. IT-49, no. 8, pp. 1912–1921, August 2003.

[14] S. Vishwanath, N. Jindal, and A. Goldsmith, "Duality, achievable rate and sum-rate capacity of Gaussian MIMO broadcast channels", *IEEE Trans. Inf. Theory*, vol. IT-49, no. 10, pp. 2659–2668, October 2003.

[15] W. Rhee, W. Yu, and J. M. Cioffi, "The optimality of beamforming in uplink multiuser wireless systems", *IEEE Trans. Wireless Commun.*, vol. TWC-3, no. 1, pp. 86–96, January 2004.

[16] N. Jindal and A. Goldsmith, "Dirty-paper coding versus TDMA for MIMO broadcast channels", in *IEEE Trans. Inf. Theory*, vol. IT-51, no. 5, pp. 1783–1794, March 2005.

[17] T. Yoo and A. Goldsmith, "On the optimality of multi-antenna broadcast scheduling using zero-forcing beamforming", *IEEE J. Select. Areas Commun.*, vol. SAC-24, no. 3, pp. 528–541, March 2006.

[18] B. Hassibi and T. L. Marzetta, "Multiple-antennas and isotropically random unitary inputs: the received signal density in closed form", *IEEE Trans. Inf. Theory*, vol. IT-48, no. 6, pp. 1473–1484, June 2002.

[19] K. K. J. Chung, C.-S. Hwang, and Y. K. Kim, "A random beamforming technique in mimo systems exploiting multiuser diversity", *IEEE J. Select. Areas Commun.*, vol. SAC-21, no. 5, pp. 848–855, June 2003.

[20] M. Sharif and B. Hassibi, "On the capacity of MIMO broadcast channels with partial side information", *IEEE Trans. Inf. Theory*, vol. IT-51, no. 2, pp. 506–522, February 2005.

Appendices

A.1 Fourier transform

Understanding the dual time-frequency nature of signals and systems is imperative to communication engineers. Fourier transforms are the essential mathematical tools for such understanding. This appendix provides a brief introduction to Fourier transforms. The Fourier transforms for both continuous-time and discrete-time signals, including the discrete Fourier transform (DFT), are discussed. Further details can be found in [1].

A.1.1 Fourier transforms for continuous-time signals

Let us first demonstrate the dual time-frequency nature of signals through a complex exponential signal given by

$$\tilde{x}(t) = Ae^{j2\pi f_0 t + \phi}, \quad -\infty < t < \infty, \tag{A.1.1}$$

which is also referred to as *rotating phasor*. A complex exponential signal is characterized by three parameters: amplitude A, frequency f_0, and phase ϕ. Thus, $\tilde{x}(t)$ can be equivalently specified in frequency domain by indicating A and ϕ values or $Ae^{j\phi}$ at frequency f_0. Applying the Euler's formula, $e^{j\omega} = \cos \omega + j \sin u$, we can show that the real part of complex exponential signal is a sinusoidal signal, i.e.,

$$\text{Re}\{\tilde{x}(t)\} = A \cos(2\pi f_0 t + \phi). \tag{A.1.2}$$

Alternatively, a sinusoidal signal can be written in terms of complex exponential signal and its complex conjugate as

$$x(t) = A \cos(2\pi f_0 t + \phi) = \frac{1}{2}(Ae^{j2\pi f_0 t + \phi} + Ae^{-j2\pi f_0 t - \phi}) = \frac{1}{2}(\tilde{x}(t) + \tilde{x}^*(t)). \tag{A.1.3}$$

Therefore, the frequency domain representation of sinusoidal signal involves specifying $A/2$ and ϕ (or $Ae^{j\phi}/2$) at frequency f_0 and $A/2$ and $-\phi$ (or $Ae^{-j\phi}/2$) at frequency $-f_0$, as shown in Figure A.1.1. Such equivalent spectral representation leads to a useful alternative perspective to signal and systems.

Fourier transform (including Fourier series) generalizes such frequency domain representation by decomposing a wide class of signals into sum of complex exponential signals. Specifically, an arbitrary energy signal $x(t)$ can be rewritten as

$$x(t) = \int_{-\infty}^{+\infty} X(f)e^{j2\pi ft}df, \tag{A.1.4}$$

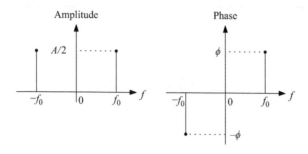

Figure A.1.1   Amplitude and phase spectra of sinusoidal signal

where $X(f)$ is the Fourier transform of $x(t)$, defined as

$$X(f) = \int_{-\infty}^{+\infty} x(t)e^{-j2\pi ft}\,dt. \tag{A.1.5}$$

We limit ourself to energy signals to ensure the existence of the integral. We use $\mathscr{F}[\cdot]$ to represent Fourier transform symbolically, i.e., $X(f) = \mathscr{F}[x(t)]$. The inverse Fourier transform is denoted by $\mathscr{F}^{-1}[\cdot]$. For periodic power signals, such decomposition is carried out using Fourier series. In particular, a periodic signal $x_p(t)$ of period T can be rewritten as

$$x_p(t) = \sum_{n=-\infty}^{+\infty} X_n e^{j2\pi nf_0 t}, \tag{A.1.6}$$

where $f_0 = 1/T$ is the fundamental frequency and X_ns are the Fourier coefficient given by

$$X_n = \frac{1}{T}\int_0^T x_p(t)e^{-j2\pi nf_0 t}\,dt \; n = \cdot, -1, 0, 1, \ldots. \tag{A.1.7}$$

For real signal $x(t)/x_p(t)$, $X(f)/X_n$ has conjugate symmetric property, i.e.,

$$|X(f)| = |X(-f)|, \quad \angle X(f) = -\angle X(-f). \tag{A.1.8}$$

The frequency domain representation based on Fourier transforms completely characterizes the energy signal. The energy of signal $x(t)$ is calculated in time domain as

$$E \triangleq \int_{-\infty}^{+\infty} |x(t)|^2\,dt. \tag{A.1.9}$$

According to the Parseval's theorem, the energy of $x(t)$ can also be calculated, using its Fourier transform $X(f)$, as

$$E = \int_{-\infty}^{+\infty} |X(f)|^2\,df. \tag{A.1.10}$$

In fact, $|X(f)|^2$ characterizes the energy distribution in the frequency domain, typically known *energy spectral density*. Similarly, the power of periodic signal $x_p(t)$, defined as

$$P \triangleq \frac{1}{T} \int_T |x_p(t)|^2 dt, \tag{A.1.11}$$

can be alternatively calculated using the Fourier coefficients as

$$P = \sum_{n=-\infty}^{+\infty} |X_n|^2. \tag{A.1.12}$$

The square of the Fourier coefficients $|X_n|^2$s quantizes the power contribution of different harmonic frequencies. For the general power signal $x(t)$ with power defined as

$$P \triangleq \lim_{T \to \infty} \frac{1}{2T} \int_{-T}^{T} |x(t)|^2 dt, \tag{A.1.13}$$

we define *power spectral density*, denoted by $S(f)$, to characterize its power distribution in frequency domain. It can be shown that $S(f)$ can be calculated as the Fourier transform of the time-average *autocorrelation function* of power signal, defined as

$$R(\tau) \triangleq \lim_{T \to \infty} \frac{1}{2T} \int_{-T}^{T} x(t)x(t+\tau)dt. \tag{A.1.14}$$

Specifically, $S(f) = \mathscr{F}[R(\tau)]$.

Fourier transform has several important properties that greatly facilitate signal and system analysis. Convolution property, probably the most important one, states that the Fourier transform of the convolution of two signals is the product of the Fourier transforms of signals, i.e.,

$$\mathscr{F}[h(t) * g(t)] = \mathscr{F}[h(t)] \cdot \mathscr{F}[g(t)] = H(f) \cdot G(f). \tag{A.1.15}$$

The multiplication property suggests that time domain multiplication implies frequency domain convolution, i.e.,

$$\mathscr{F}[h(t)g(t)] = \mathscr{F}[h(t)] * \mathscr{F}[g(t)] = \int_{-\infty}^{+\infty} H(\lambda)G(f-\lambda)d\lambda. \tag{A.1.16}$$

Frequency domain shift property suggests that multiplying a complex exponential signal to a signal will shift its spectrum in frequency domain, i.e.,

$$\mathscr{F}[x(t)e^{j2\pi f_0 t}] = X(f - f_0). \tag{A.1.17}$$

A.1.2 Fourier transform for discrete-time signal

Similarly, the Fourier transform decomposes discrete-time signals into discrete-time complex exponential signals, given by

$$\tilde{x}[n] = e^{j2\pi \omega_0 n}, \quad -\infty < n < \infty, \tag{A.1.18}$$

where n is the integer index, and ω_0 is the frequency. Note that two discrete-time complex exponential signals are identical if their frequencies differ by an integer multiple of 2π. As such, the frequency range of interest for discrete-time signal is $[-\pi, \pi]$. A discrete-time energy signal $x[n]$ can be decomposed as

$$x[n] = \frac{1}{2\pi} \int_{-\pi}^{\pi} X(\omega) e^{j\omega n} d\omega, \tag{A.1.19}$$

where $X(\omega)$ is the discrete-time Fourier transform (DTFT) of $x[n]$, given by

$$X(\omega) = \sum_{n=-\infty}^{+\infty} x[n] e^{-j\omega n}. \tag{A.1.20}$$

The convergence of the summation is assured under the energy signal assumption. $X(\omega)$ is periodic in ω with period of 2π. As such, the period of $X(\omega)$ over $[-\pi, \pi]$ contains all the information of $x[n]$. The energy of $x[n]$, defined as

$$E \triangleq \sum_{n=-\infty}^{+\infty} |x[n]|^2 \tag{A.1.21}$$

can be alternatively calculated using $X(\omega)$ as

$$E = \frac{1}{2\pi} \int_{-\pi}^{\pi} |X(\omega)|^2 d\omega. \tag{A.1.22}$$

$X(\omega)$ is also conjugate symmetric for real signal $x[n]$.

$X(\omega)$ is continuous function of ω. There are uncountable infinite number of samples within a period of $[-\pi, \pi]$. Such alternative representation of $x[n]$ is inefficient, especially when $x[n]$ is of finite length. DFT leads to a more efficient equivalent frequency domain representation of finite-length discrete-time signal. In particular, a discrete-time signal of length L, $x[n]$ can be rewritten in terms of N, $N \geq L$, *harmonically related* complex exponentials $e^{j(2\pi k/N) \cdot n}$, $k = 0, 1, \ldots, N - 1$, as

$$x[n] = \frac{1}{N} \sum_{k=0}^{N-1} X[k] e^{j(2\pi k/N) \cdot n}, \quad n = 0, 1, \ldots, L. \tag{A.1.23}$$

Here $X[k]$ is the N-point DFT of $x[n]$, given by

$$X[k] = \sum_{n=0}^{L-1} x[n] e^{-j(2\pi k/N) \cdot n}, \quad k = 0, 1, \ldots, N - 1. \tag{A.1.24}$$

Let us define the constant $W_N = e^{-j2\pi/N}$. The DFT and its inverse operation, inverse DFT (IDFT), can be rewritten as

$$X[k] = \sum_{n=0}^{L-1} x[n] W_N^{k \cdot n} \quad \text{and} \quad x[n] = \frac{1}{N} \sum_{k=0}^{N-1} X[k] W_N^{-k \cdot n}. \tag{A.1.25}$$

Both DFT and IDFT can be efficiently calculated using the fast Fourier transform (FFT) algorithm.

DTFT shares the similar properties of continuous-time Fourier transform. Meanwhile, DFT possesses some unique properties as DFT is defined on finite-length discrete-time signal. In particular, the time-shift property of DTFT states

$$\text{DTFT}\{x[n - n_d]\} = e^{-j\omega n_d}X(\omega), \tag{A.1.26}$$

where n_d is the amount of shift applied to $x[n]$ in time domain. On the other hand, the time-shift property of N-point DFT implies

$$\text{DFT}\{x[(n - n_d)_N]\} = W_N^{k \cdot n_d}X[k], \tag{A.1.27}$$

where $(\cdot)_N$ denotes modulo-N operation. $x[(n - n_d)_N]$ is the circularly shift version of $x[n]$ by n_d samples with period of N. For example, if $x[n] = [1, 2, 4]$, then $x[(n - 2)_4] = [4, 0, 1, 2]$. The convolution property of DTFT states

$$\text{DTFT}\{x[n] * y[n]\} = \text{DTFT}\{x[n]\} \cdot \text{DTFT}\{y[n]\} = X(\omega) \cdot Y(\omega), \tag{A.1.28}$$

where $*$ denotes the linear convolution of two sequences, defined by

$$x[n] * y[n] = \sum_{k=-\infty}^{+\infty} x[k] \cdot y[n - k]. \tag{A.1.29}$$

The convolution property of DFT applies to the circular convolution operation, defined by

$$x[n] \circledast y[n] = \sum_{m=0}^{N-1} x[m]y[(n - m)_N]. \tag{A.1.30}$$

Here the lengths of $x[n]$ and $y[n]$ should be no greater than the period of circular shift N. The N-point DFT of the circular convolution result is

$$\text{DFT}\{x[n] \circledast y[n]\} = \text{DFT}\{x[n]\} \cdot \text{DFT}\{y[n]\} = X[k] \cdot Y[k]. \tag{A.1.31}$$

Circular convolution renders the same result of linear convolution after zero padding the signals to sufficient length. As such, FFT algorithms can apply to calculate linear convolution more efficiently than direct time-domain approach. Meanwhile, linear convolution can generate circular convolution results with cyclic prefix, as shown in Chapter 6 when we discuss OFDM transmission.

A.2 Probability and random variables

This appendix provides a brief introduction to probability and random variables. Some basic concepts about random processes are also presented. Further details can be found in [2].

A.2.1   *Probabilistic modeling with random variables*

The concepts of probability and random variables are introduced to characterize events whose outcome is not known deterministically before it happens. Example includes tomorrow's temperature, stock prices, and noise voltage at a future time instant, etc. While the future outcome of these events is not deterministically known, some historical data and experience about these events are usually available. For example, experience shows that the noise voltage will take values around zero. The variation around zero is larger when the room temperature is higher. To explore these previous knowledge and experience of such random events, we generally adopt a probabilistic approach. Specifically, the future outcome of such event is modeled as a *random variable*, and previous knowledge about the event is captured with its distribution functions.

If the possible outcomes are finite or countable infinite, the corresponding random variable is called *discrete random variable*. The associated distribution function is termed as *probability mass function* (PMF). Let X be a discrete random variable with possible values $x_1, x_2, \ldots, x_K$. The PMF of X is given by

$$P_X(x_i) = \Pr[X = x_i], \quad i = 1, 2, \ldots, K. \tag{A.2.1}$$

The PMF should satisfy that $0 \le P_X(x_i) \le 1$ for every i and $\sum_i P_X(x_i) = 1$. For example, let us consider the event of rolling a six-sided die. If the die is fair, then the probability of each of the six possible numbers will be equal to 1/6. The modeling random variable X will be discrete with six possible values, $1, 2, 3, 4, 5,$ and 6. The corresponding PMF is

$$P_X(i) = \Pr[X = i] = 1/6, \quad i = 1, 2, \ldots, 6. \tag{A.2.2}$$

Continuous random variable is used to model events with uncountable infinite outcomes, i.e., with possible outcome value over a continuous region. The *probability density function* (PDF) is the distribution function associated with a continuous random variable. The PDF of random variable X, denoted by $p_X(\cdot)$, is a nonnegative valued function defined over the value range of the random variable, i.e., $p_X(x) \ge 0$ for all x. The value of PDF at a certain value x_0 signifies the likelihood of the random variable taking values around x_0. The PDF should satisfy $\int_{-\infty}^{+\infty} p_X(x)dx = 1$. For example, experience shows the noise voltage in a circuit typically takes positive/negative values around zero. The histogram of noise voltage measurements has a symmetric bell shape. These observations motivate the modeling of noise voltage with a Gaussian random variable. The PDF of a Gaussian random variable is given by

$$p(x) = \frac{1}{\sqrt{2\pi}\sigma} \exp\left(-\frac{(x-\mu)^2}{2\sigma^2}\right), \tag{A.2.3}$$

which has a symmetric bell shape, as shown in Figure A.2.1. We can match the noise voltage histogram well with properly chosen values for μ and σ.

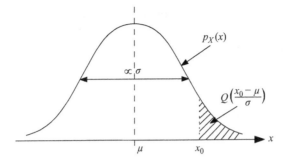

Figure A.2.1   PDF of Gaussian random variable

A.2.2   Properties of random variables

The *expected value* or *mean* characterizes the probabilistic average of the possible outcomes of a random variable. The mean of continuous random variable X is defined as

$$\mu_X \stackrel{\Delta}{=} \mathbf{E}[X] = \int_{-\infty}^{+\infty} x p_X(x) dx. \tag{A.2.4}$$

The integration becomes summation for discrete random variables. The nth moment of random variable X is defined as

$$\mathbf{E}[X^n] = \int_{-\infty}^{+\infty} x^n p_X(x) dx. \tag{A.2.5}$$

Apparently, the mean is the same as the first moment. The *variance* of a random variable characterizes the deviation of the possible outcome values from its mean. Mathematically, it is defined as

$$\sigma_X^2 \stackrel{\Delta}{=} \text{Var}[X] = \mathbf{E}[(X - \mu_X)^2] = \mathbf{E}[X^2] - (\mathbf{E}[X])^2. \tag{A.2.6}$$

Variance is also known as the second central moment of the random variable. It can be shown that parameters μ and σ^2 in the PDF of a Gaussian random variable are its mean and variance, respectively.

With the PDF $p_X(x)$, we can calculate the probability that the random variable X takes values in a certain region. Specifically, the probability that the realization of X is smaller than a constant value x is calculated as

$$\Pr[X \le x] = \int_{-\infty}^{x} p_X(z) dz \stackrel{\Delta}{=} F_X(x), \tag{A.2.7}$$

where $F_X(\cdot)$ is known as the *cumulative distribution function* (CDF) of random variable X. The probability that X falls into the interval $[x_1, x_2]$ is given by

$$\Pr[x_1 \le X \le x_2] = \int_{x_1}^{x_2} p_X(z) dz = F_X(x_2) - F_X(x_1). \tag{A.2.8}$$

As an example, the CDF of a Gaussian random variable is given by

$$F_X(x) = \frac{1}{\sqrt{2\pi}\sigma} \int_{-\infty}^{x} \exp\left(-\frac{(z-\mu)^2}{2\sigma^2}\right) dz. \tag{A.2.9}$$

The integral cannot be solved into closed form.

Since the Gaussian CDF is frequently used in many areas, a special function is defined and tabulated to facilitate its evaluation. More specifically, the CDF of Gaussian random variable can be written as

$$F_X(x) = 1 - Q\left(\frac{x-\mu}{\sigma}\right), \tag{A.2.10}$$

where $Q(\cdot)$ is the *Gaussian Q-function*, defined by

$$Q(x) = \frac{1}{\sqrt{2\pi}} \int_{x}^{\infty} \exp\left(-\frac{z^2}{2}\right) dz. \tag{A.2.11}$$

$Q(x)$ gives the probability that a Gaussian random variable with zero mean and unit variance is larger than x. Sample values of the Q-function is given in Table A.2.1 [3]. The Gaussian Q-function is related to complementary error function as $Q(x) = \text{erfc}(x/\sqrt{2})/2$.

Given a random variable X with PDF $p_X(x)$, let Y denote a function of X, i.e., $Y = g(X)$, where $g(\cdot)$ is a function defined on the real line. Y becomes another random variable. If $g(\cdot)$ is a one-to-one function, then the PDF of Y is given by

$$p_Y(y) = \frac{p_X(x)}{|g'(x)|}\Big|_{x=g^{-1}(y)}, \tag{A.2.12}$$

where $g'(\cdot)$ denotes the derivative and $g^{-1}(\cdot)$ the inverse function of function $g(\cdot)$. The mean and variance of Y can be calculated with the resulting PDF. Alternatively, we can directly calculate the mean of $Y = g(X)$ as

$$\mu_Y \triangleq \mathbf{E}[g(X)] = \int_{-\infty}^{+\infty} g(x)p_X(x)dx. \tag{A.2.13}$$

The *moment generating function* (MGF) is another distribution function that characterizes a random variable. The MGF of random variable X is defined as

$$\mathcal{M}_X(s) = \int_{-\infty}^{+\infty} p_X(x)e^{s \cdot x}dx. \tag{A.2.14}$$

where s is a complex dummy variable and $p_X(\cdot)$ is the PDF of X. Note that the MGF of X is related to the Laplace transform of the PDF as $p_X(\cdot)$ as $\mathcal{M}_X(-s) = \mathcal{L}\{p_X(x)\}$. MGF is so named because it allows for easy calculation of the moments. Specifically, the nth moment of random variable X can be calculated as

$$\mathbf{E}[X^n] = \frac{d^n}{ds^n} \mathcal{M}_X(s)\Big|_{s=0}. \tag{A.2.15}$$

Table A.2.1   Sample values of Gaussian Q-function

x	$Q(x)$	x	$Q(x)$	x	$Q(x)$
0.00	0.5	2.00	0.022750	4.00	3.1671×10^{-5}
0.05	0.48006	2.05	0.020182	4.05	2.5609×10^{-5}
0.10	0.46017	2.10	0.017864	4.10	2.0658×10^{-5}
0.15	0.44038	2.15	0.015778	4.15	1.6624×10^{-5}
0.20	0.42074	2.20	0.013903	4.20	1.3346×10^{-5}
0.25	0.40129	2.25	0.012224	4.25	1.0689×10^{-5}
0.30	0.38209	2.30	0.010724	4.30	8.5399×10^{-6}
0.35	0.36317	2.35	0.0093867	4.35	6.8069×10^{-6}
0.40	0.34458	2.40	0.0081975	4.40	5.4125×10^{-6}
0.45	0.32636	2.45	0.0071428	4.45	4.2935×10^{-6}
0.50	0.30854	2.50	0.0062097	4.50	3.3977×10^{-6}
0.55	0.29116	2.55	0.0053861	4.55	2.6823×10^{-6}
0.60	0.27425	2.60	0.0046612	4.60	2.1125×10^{-6}
0.65	0.25785	2.65	0.0040246	4.65	1.6597×10^{-6}
0.70	0.24196	2.70	0.0034670	4.70	1.3008×10^{-6}
0.75	0.22663	2.75	0.0029798	4.75	1.0171×10^{-6}
0.80	0.21186	2.80	0.0025551	4.80	7.9333×10^{-7}
0.85	0.19766	2.85	0.0021860	4.85	6.1731×10^{-7}
0.90	0.18406	2.90	0.0018658	4.90	4.7918×10^{-7}
0.95	0.17106	2.95	0.0015889	4.95	3.7107×10^{-7}
1.00	0.15866	3.00	0.0013499	5.00	2.8665×10^{-7}
1.05	0.14686	3.05	0.0011442	5.05	2.2091×10^{-7}
1.10	0.13567	3.10	0.0009676	5.10	1.6983×10^{-7}
1.15	0.12507	3.15	0.00081635	5.15	1.3024×10^{-7}
1.20	0.11507	3.20	0.00068714	5.20	9.9644×10^{-8}
1.25	0.10565	3.25	0.00057703	5.25	7.605×10^{-8}
1.30	0.09680	3.30	0.00048342	5.30	5.7901×10^{-8}
1.35	0.088508	3.35	0.00040406	5.35	4.3977×10^{-8}
1.40	0.080757	3.40	0.00033693	5.40	3.332×10^{-8}
1.45	0.073529	3.45	0.00028029	5.45	2.5185×10^{-8}
1.50	0.066807	3.50	0.00023263	5.50	1.899×10^{-8}
1.55	0.060571	3.55	0.00019262	5.55	1.4283×10^{-8}
1.60	0.054799	3.60	0.00015911	5.60	1.0718×10^{-8}
1.65	0.049471	3.65	0.00013112	5.65	8.0224×10^{-9}
1.70	0.044565	3.70	0.00010780	5.70	5.9904×10^{-9}
1.75	0.040059	3.75	8.8417×10^{-5}	5.75	4.4622×10^{-9}
1.80	0.035930	3.80	7.2348×10^{-5}	5.80	3.3157×10^{-9}
1.85	0.032157	3.85	5.9059×10^{-5}	5.85	2.4579×10^{-9}
1.90	0.028717	3.90	4.8096×10^{-5}	5.90	1.8175×10^{-9}
1.95	0.025588	3.95	3.9076×10^{-5}	5.95	1.3407×10^{-9}

A.2.3   Sample random variables

Binomial random variable characterizes the number of successes in a series of independent trials. Let us assume that there are N independent trials, each with a success probability of p, then the number of successes, X, follows binomial distribution with PMF given by

$$\Pr[X = n] = \binom{N}{n} p^n (1 - p)^{N-n}, \quad n = 0, 1, 2, \ldots, N. \tag{A.2.16}$$

The mean of binomial random variable is equal to Np and the variance equal to $Np(1 - p)$.

Geometric random variable models the number of trials until the first success. The PMF of geometric random variable is given by

$$\Pr[X = k] = p(1 - p)^k, \quad k = 0, 1, 2, \ldots, \tag{A.2.17}$$

where p is the success probability. Geometric random variable has a mean of $1/(1 - p)$ and variance of $p/(1 - p)^2$.

Uniform random variable characterizes random events that return value with a final interval with equal likelihood. The PDF of a uniform random variable X defined over the interval $[a, b]$ is given by

$$p_X(x) = \frac{1}{b - a}, \quad a \leq x \leq b. \tag{A.2.18}$$

The mean of uniform random variable is equal to $(a + b)/2$ and the variance of X equal to $(b - a)^2/12$.

Exponential random variable is widely used in engineering and life science. The PDF of an exponential random variable X is given by

$$p_X(x) = \frac{1}{\mu} e^{-x/\mu}, \quad x \geq 0. \tag{A.2.19}$$

Exponential random variable has a mean of μ and variance of μ^2. The MGF of an exponential random variable can be shown to be given by

$$\mathcal{M}_X(s) = \frac{1}{1 - s\mu}. \tag{A.2.20}$$

A.2.4   Multiple random variables

Multiple random variables are used to study multiple random events. Previous knowledge about these events is captured with the *joint PDF*. Let X and Y denote two random variables, with joint PDF denoted by $p_{X,Y}(x, y)$. The value of $p_{X,Y}(x, y)$ at (x_0, y_0) indicates the likelihood of X taking value around x_0 and at the same time Y taking values around y_0. The joint PDF should satisfy $p_{X,Y}(x, y) \geq 0$ for all x and y and $\int_{-\infty}^{+\infty} \int_{-\infty}^{+\infty} p_{X,Y}(x, y) dx dy = 1$. From the joint PDF, we can obtain the PDF of X as

$$p_X(x) = \int_{-\infty}^{+\infty} p_{X,Y}(x, y) dy, \tag{A.2.21}$$

which is also called the *marginal PDF*. The marginal PDF of random Y can be similar obtained.

Random events will be dependent if the observation of one event affects the likelihood of other event's outcomes. The dependence between random events can be studied through the joint distribution function of the corresponding random variables. The likelihood of random variable Y given the realization x for random variable X is characterized by the *conditional PDF*, denoted by $p_{Y|X=x}(y)$. If $p_{Y|X=x}(y)$ remains the same for different x values, then X and Y are independent of each other. It can be shown that conditional PDF can be calculated as

$$p_{Y|X=x}(y) = \frac{p_{X,Y}(x,y)}{p_X(x)}. \tag{A.2.22}$$

X and Y are independent if the joint PDF $p_{X,Y}(x,y)$ can be written as the product of the marginal PDFs as $p_{X,Y}(x,y)=p_X(x)p_Y(y)$, which leads to $p_{Y|X=x}(y)=p_Y(y)$. The joint probability of two independent random events can be calculated as the product of the probabilities of individual events. In particular, when X and Y are independent continuous random variables, the joint probability of $X < x$ and $Y < y$ is given by

$$\Pr[X < x, Y < y] = \Pr[X < x] \cdot \Pr[Y < y]. \tag{A.2.23}$$

For dependent random events, the marginal probability of one event can be calculated by conditioning on the outcomes of other dependent events. More specifically, consider two dependent random variables X and Y. The probability of $X < x$ can be calculated as

$$\Pr[X < x] = \sum_{i=1}^{N} \Pr[X < x, Y = y_i] = \sum_{i=1}^{N} \Pr[X < x|Y = y_i] \Pr[Y = y_i], \tag{A.2.24}$$

when Y is a discrete random variable, with possible outcomes y_i, $i = 1, 2, \ldots, N$. Here, $\Pr[X < x|Y = y_i]$ denotes the conditional probability of $X < x$ given $Y = y_i$. Such relationship is often referred to as the *total probability theorem*. When Y is a continuous random variable, the relationship becomes

$$\Pr[X < x] = \int_{y=-\infty}^{+\infty} \Pr[X < x|Y = y] p_Y(y) dy. \tag{A.2.25}$$

The dependence of two random events is characterized by the *correlation coefficient* between the corresponding random variables. The correlation coefficient of random variables X and Y is defined as

$$\rho = \frac{E[XY] - \mu_X \mu_Y}{\sigma_X \sigma_Y}, \tag{A.2.26}$$

where $E[XY]$ is the correlation of X and Y. If two random variables are independent, their correlation coefficient will be zero. On the other hand, zero correlation coefficient implies that two random variables are uncorrelated but not necessarily independent, except for Gaussian random variables. Meanwhile, a correlation coefficient close to one indicates that the two random events are highly dependent.

Let X_1 and X_2 be two continuous random variables with joint PDF $p_{X_1,X_2}(x_1,x_2)$. Consider two continuous and differentiable functions of these variables $Y_1 = h_1(X_1,X_2)$ and $Y_2 = h_2(X_1,X_2)$. The joint PDF of Y_1 and Y_2 is given by

$$p_{Y_1,Y_2}(y_1,y_2) = |J| \cdot p_{X_1,X_2}(x_1,x_2)|_{x_1=g_1(y_1,y_2),x_2=g_2(y_1,y_2)}, \qquad (A.2.27)$$

where $g_1(\cdot,\cdot)$ and $g_2(\cdot,\cdot)$ represent the inverse relationship of X_1 and X_2 in terms of Y_2 and Y_2, respectively, and J is the Jacobian of the inverse relationship, defined as

$$J = \det \begin{bmatrix} \frac{\partial g_1(y_1,y_2)}{\partial y_1} & \frac{\partial g_1(y_1,y_2)}{\partial y_2} \\ \frac{\partial g_2(y_1,y_2)}{\partial y_1} & \frac{\partial g_2(y_1,y_2)}{\partial y_2} \end{bmatrix}. \qquad (A.2.28)$$

In certain applications, we are interested in the distribution of the *sum of multiple random variables*. The PDF of the sum of two continuous random variables X and Y, denoted by Z, i.e., $Z = X + Y$, is calculated as

$$p_Z(z) = \int_{-\infty}^{+\infty} p_{X,Y}(z - u, u)du, \qquad (A.2.29)$$

where $p_{X,Y}(\cdot,\cdot)$ denotes the joint PDF of X and Y. If X and Y are independent random variables, $p_Z(z)$ is given by

$$p_Z(z) = \int_{-\infty}^{+\infty} p_X(z - u)p_Y(u)du, \qquad (A.2.30)$$

which involves the convolution of the PDFs of X and Y. The statistics of the sum of independent random variables can be more conveniently obtained using the MGF approach. In particular, the MGF of $Z = X + Y$ can be shown to be equal to

$$\mathcal{M}_Z(s) = \mathcal{M}_X(s) \cdot \mathcal{M}_Y(s), \qquad (A.2.31)$$

i.e., the MGF of the sum is equal to the product of individual MGFs for independent random variables.

The Gaussian random variable is widely used in communication engineering, partly because of the *central-limit theorem*, which states that the sum of multiple independent random variables can be well approximated by a Gaussian random variable. Formally, let $X_1, X_2, \ldots, X_N$ be N independent and identically distributed random variables, each with finite mean μ and finite variance σ^2. Their sum, denoted by $Z = \sum_{n=1}^{N} X_n$ is approximately Gaussian distributed with mean $N\mu$ and variance $N\sigma^2$. The larger the N value, the better the approximation. In some cases, the independent and identically distributed assumption can be relaxed, but none of individual random variable should dominate the sum.

A.2.5 Random processes

Random processes model random events that evolve over time or space, e.g., noise voltage over a symbol period. We denote a random process by $X(t)$ and its realization by $x(t)$. The sample of $X(t)$ at a particular time instant is a random variable, which may be correlated to the sample $X(t)$ at another time instant. The correlation of two samples of random process $X(t)$ is characterized by the *autocorrelation function*, defined as

$$\rho_X(t, s) = \mathbf{E}[X(t) \cdot X(s)]. \tag{A.2.32}$$

A random process $X(t)$ is *stationary* if the marginal and joint distributions of its arbitrary number of samples remain constant over time. In particular, the mean of the sample of $X(t)$ is a constant over time, i.e., $\mathbf{E}[X(t)] = \mu_X$. Furthermore, the autocorrelation function $\rho_X(t, s)$ is dependent only on the time difference of samples $t - s$, i.e., $\rho_X(t, s) = \rho_X(t, t + \tau) = \rho_X(\tau)$. In general, it is very difficult to verify the stationarity of a random process against its formal definition. Typically, we can only confirm the stationary behavior of the mean and autocorrelation for a random process. In this case, we call the process *wide-sense stationary*.

A random process $X(t)$ is ergodic if the time average of its realization $x(t)$ is equal to the corresponding statistical average at a particular time instant, i.e.,

$$\mathbf{E}[X(t_0)] = \lim_{t \to \infty} \frac{1}{T} \int_{-T/2}^{T/2} x(t) dt. \tag{A.2.33}$$

Note that the right hand side is the time average of a particular realization, and the left hand side is the statistical average at t_0. In many applications, stationary processes are assumed to be ergodic.

The realization of a stationary random process is in general a power signal. The power of stationary random process $X(t)$ can be calculated as

$$P = \mathbf{E}[X^2(t)] = \rho_X(0). \tag{A.2.34}$$

The power distribution in the frequency domain of all realizations of the same random process $X(t)$ are the same and commonly characterized by *power spectral density*, denoted by $S_X(f)$. It has been established by the *Wiener–Khintchine theorem* that the power spectral density, and the autocorrelation function of a stationary random process form a Fourier transform pair. Mathematically, we have

$$S_X(f) = \mathscr{F}[\rho_X(\tau)] = \int_{-\infty}^{+\infty} \rho_X(\tau) e^{-j2\pi f \tau} d\tau, \tag{A.2.35}$$

and

$$\rho_X(\tau) = \mathscr{F}^{-1}[\rho_X(\tau)] = \int_{-\infty}^{+\infty} S_X(f) e^{j2\pi f \tau} df. \tag{A.2.36}$$

Setting τ to zero in the above inverse Fourier transform formula, we have

$$\rho_X(0) = \int_{-\infty}^{+\infty} S_X(f) df = P, \tag{A.2.37}$$

which partially confirms that $S_X(f)$ gives the power distribution of $X(t)$ in frequency domain.

As an example, let us consider a stationary white Gaussian process $W(t)$, typically used to model noise in communication systems. The word "Gaussian" in the name indicates that the sample of $W(t)$ at any time instant is a Gaussian random variable. The word "stationary" suggests that the mean μ_W and variance σ_W^2 of the Gaussian random variables at different time instances remain the same, and the autocorrelation function of $W(t)$ is only a function of the time difference τ. The word "white" implies that the samples with nonzero time difference are uncorrelated, i.e., the autocorrelation function $\rho_W(\tau)$ is an impulse function given by

$$\rho_W(\tau) = \frac{N_0}{2}\delta(\tau), \tag{A.2.38}$$

where $\delta(\cdot)$ denotes the Dirac impulse function. Taking Fourier transform of the autocorrelation function, we obtain the power spectral density function as

$$S_W(f) = \frac{N_0}{2}, \tag{A.2.39}$$

which remains constant for all f. In other words, the power of white process is equally distributed over all frequency components. N_0 is usually referred to as noise spectral density.

A.3 Vectors and matrices

This appendix provides a brief introduction to vectors and matrices, including their properties and operations. The singular value decomposition (SVD) of arbitrary matrix is also discussed [4].

A.3.1 Definition

A *vector* typically refers to a column of numbers. The number of rows of a vector is called its *dimension*. An N-dimensional vector $\mathbf{x}$ is given by

$$\mathbf{x} = \begin{bmatrix} x_1 \\ x_2 \\ \vdots \\ x_N \end{bmatrix}. \tag{A.3.1}$$

The ith entry x_i is a scalar. A *matrix* is in general a rectangular array of numbers. A matrix with N rows and M columns is usually referred to as an $N \times M$ matrix or an N-by-M matrix, given by

$$\mathbf{A} = \begin{bmatrix} a_{11} & a_{12} & \cdots & a_{1M} \\ a_{21} & a_{22} & \cdots & a_{2M} \\ \vdots & \vdots & \ddots & \vdots \\ a_{N1} & a_{N2} & \cdots & a_{NM} \end{bmatrix}. \tag{A.3.2}$$

Scalar a_{ij} denotes its entry at ith row and jth column. When $M = N$, we have a square matrix. Clearly, a vector is a special matrix with $M = 1$. We can also have a row vector when $N = 1$.

The diagonal elements of matrix **A** are those entries a_{ij} with $i = j$. A diagonal matrix is a special square matrix whose off-diagonal entries, i.e., a_{ij} with $i \neq j$, are all zeros. The identity matrix $\mathbf{I}_N$ is an $N \times N$ diagonal matrix whose diagonal entries are all ones, i.e.,

$$
\mathbf{I}_N = \begin{bmatrix} 1 & 0 & \cdots & 0 \\ 0 & 1 & \cdots & 0 \\ \vdots & \vdots & \ddots & \vdots \\ 0 & 0 & \cdots & 1 \end{bmatrix}.
$$

(A.3.3)

The columns of $\mathbf{I}_N$, called *unit vectors*, are special vectors with one entry equal to one and the rest equal to 0.

Several metrics are defined for vectors and matrices. The Euclidean norm, or simply the norm, of vector **x** is defined as

$$
\|\mathbf{x}\| = \sqrt{\sum_{i=1}^{N} |x_i|^2}.
$$

(A.3.4)

The corresponding norm for matrices are called Frobenius norm, which is defined for an N-by-M matrix **A** as

$$
\|\mathbf{A}\| = \sqrt{\sum_{i=1}^{N} \sum_{j=1}^{M} |a_{ij}|^2}.
$$

(A.3.5)

The *trace* of an N-by-N square matrix **A** is the sum of its diagonal elements, defined as

$$
\mathrm{Tr}[\mathbf{A}] = \sum_{i=1}^{N} a_{ii}.
$$

(A.3.6)

Several rows of a matrix form a linearly independent set if none of the rows can be represented as a linear combination of the remaining rows in the set. The rank of a matrix **A**, denoted by $R_\mathbf{A}$, is equal to the number of rows in the largest linearly independent subset of its rows. It can be shown that $R_\mathbf{A}$ is also equal to the number of columns in the largest linearly independent subset of its columns. Therefore, $R_\mathbf{A} \leq \min[N, M]$. If $R_\mathbf{A} = \min[N, M]$, the matrix **A** is of full rank.

The determinant of an N-by-N square matrix **A** is defined as

$$
\det[\mathbf{A}] = \sum_{i=1}^{N} a_{ij} c_{ij}, \ 1 \leq j \leq N,
$$

(A.3.7)

where c_{ij} is the cofactor of entry a_{ij} defined by

$$
c_{ij} = (-1)^{i+j} \det[\mathbf{A}'_{ij}].
$$

(A.3.8)

Here $\mathbf{A}'_{ij}$ is the submatrix of $\mathbf{A}$ with the row and column involving a_{ij} removed. For example, $\mathbf{A}'_{12}$ obtained from $\mathbf{A}$ as

$$\mathbf{A} = \begin{bmatrix} a_{11} & a_{12} & \cdots & a_{1N} \\ a_{21} & a_{22} & \cdots & a_{2N} \\ \vdots & \vdots & \ddots & \vdots \\ a_{N1} & a_{N2} & \cdots & a_{NN} \end{bmatrix} \Rightarrow \mathbf{A}'_{12} = \begin{bmatrix} a_{12} & a_{13} & \cdots & a_{1N} \\ a_{32} & a_{32} & \cdots & a_{3N} \\ \vdots & \vdots & \ddots & \vdots \\ a_{N2} & a_{N3} & \cdots & a_{NN}, \end{bmatrix} \tag{A.3.9}$$

which is of size $(N-1) \times (N-1)$. The determinant of a 2×2 matrix $\mathbf{A}$ is given by

$$\det[\mathbf{A}] = \det \begin{bmatrix} a_{11} & a_{12} \\ a_{21} & a_{22} \end{bmatrix} = a_{11}a_{22} - a_{21}a_{12}. \tag{A.3.10}$$

A.3.2 Operations

The transpose of an N-by-M matrix $\mathbf{A}$, denoted by $\mathbf{A}^T$, is given by

$$\mathbf{A}^T = \begin{bmatrix} a_{11} & a_{12} & \cdots & a_{1M} \\ a_{21} & a_{22} & \cdots & a_{2M} \\ \vdots & \vdots & \ddots & \vdots \\ a_{N1} & a_{N2} & \cdots & a_{NM} \end{bmatrix}^T = \begin{bmatrix} a_{11} & a_{21} & \cdots & a_{M1} \\ a_{12} & a_{22} & \cdots & a_{M2} \\ \vdots & \vdots & \ddots & \vdots \\ a_{1N} & a_{2N} & \cdots & a_{MN} \end{bmatrix}. \tag{A.3.11}$$

Essentially, the ith row of $\mathbf{A}$ becomes the ith column of $\mathbf{A}^T$, leading $\mathbf{A}^T$ to be an M-by-N matrix. The transpose operation on a vector is a special case, where the transpose of a column vector leads to a row vector with the same entries. The complex conjugate of a matrix/vector $\mathbf{A}^*/\mathbf{x}^*$ is obtained by taking the complex conjugate of each entry. The conjugate transpose, also known as the Hermitian, of a matrix $\mathbf{A}$, is defined by $\mathbf{A}^H = (\mathbf{A}^*)^T$. Specifically,

$$\mathbf{A}^H = \begin{bmatrix} a_{11} & a_{12} & \cdots & a_{1M} \\ a_{21} & a_{22} & \cdots & a_{2M} \\ \vdots & \vdots & \ddots & \vdots \\ a_{N1} & a_{N2} & \cdots & a_{NM} \end{bmatrix}^H = \begin{bmatrix} a_{11}^* & a_{21}^* & \cdots & a_{M1}^* \\ a_{12}^* & a_{22}^* & \cdots & a_{M2}^* \\ \vdots & \vdots & \ddots & \vdots \\ a_{1N}^* & a_{2N}^* & \cdots & a_{MN}^* \end{bmatrix}. \tag{A.3.12}$$

Linear operations, including scalar multiplication, addition, and subtraction, on vectors/matrices essentially mean entry-by-entry operation. Therefore, the involved vector/matrices should have the same dimension. Matrices can be multiplied together only if the involved matrices have compatible dimensions. In particular, matrix $\mathbf{A}$ can be multiplied to matrix $\mathbf{B}$ only if the number of columns of $\mathbf{A}$ is equal to the number of rows of **B**, i.e., if **A** is $N \times M$, then **B** should be $M \times L$, where N and L can be any integers. The multiplication result of $\mathbf{A} \cdot \mathbf{B}$ will be an $N \times L$ matrix given by

$$\mathbf{A} \cdot \mathbf{B} = \begin{bmatrix} a_{11} & a_{12} & \cdots & a_{1M} \\ a_{21} & a_{22} & \cdots & a_{2M} \\ \vdots & \vdots & \ddots & \vdots \\ a_{N1} & a_{N2} & \cdots & a_{NM} \end{bmatrix} \begin{bmatrix} b_{11} & b_{12} & \cdots & b_{1L} \\ b_{21} & b_{22} & \cdots & b_{2L} \\ \vdots & \vdots & \ddots & \vdots \\ b_{M1} & b_{M2} & \cdots & b_{ML} \end{bmatrix} \tag{A.3.13}$$

$$= \begin{bmatrix} \sum_{j=1}^{M} a_{1j}b_{j1} & \sum_{j=1}^{M} a_{1j}b_{j2} & \cdots & \sum_{j=1}^{M} a_{1j}b_{jL} \\ \sum_{j=1}^{M} a_{2j}b_{j1} & \sum_{j=1}^{M} a_{2j}b_{j1} & \cdots & \sum_{j=1}^{M} a_{2j}b_{jL} \\ \vdots & \vdots & \ddots & \vdots \\ \sum_{j=1}^{M} a_{Nj}b_{j1} & \sum_{j=1}^{M} a_{Nj}b_{j2} & \cdots & \sum_{j=1}^{M} a_{Nj}b_{jL} \end{bmatrix}.$$

The associative and distributive laws apply to matrix multiplication as long as the dimensions are compatible. Matrix multiplication is not commutative in general. Note that $\mathbf{B} \cdot \mathbf{A}$ exists only if $N = L$. Even if $N = M = L$, $\mathbf{A} \cdot \mathbf{B}$ may not equal to $\mathbf{B} \cdot \mathbf{A}$. Square matrix can be multiplied to itself. In particular, if **A** is an $N \times N$ matrix, then $\mathbf{A}^2 = A \cdot A$. An $N \times M$ matrix can be multiplied to a column vector of size M, which leads to a column vector of size N. When a row vector is multiplied to a column vector of the same size, the result is a scalar. For example, the norm of vector **x** can be calculated as

$$\|\mathbf{x}\| = \sqrt{\mathbf{x}^H \mathbf{x}}. \tag{A.3.14}$$

If the multiplication of two square matrices leads to an identity matrix, then they are the inverse of each other, i.e., if $\mathbf{B} \cdot \mathbf{A} = \mathbf{I}_N$, then **B** is the inverse of **A**, denoted by $\mathbf{A}^{-1}$. Note that $\mathbf{A} \cdot \mathbf{A}^{-1}$ is also equal to $\mathbf{I}_N$. A square matrix is invertible if and only the matrix is of full rank. Noninvertible matrix is also called *singular matrix*. When the inverse of a square matrix is equal to its Hermitian, i.e., $\mathbf{U}^{-1} = \mathbf{U}^H$, we have $\mathbf{U}^H \mathbf{U} = \mathbf{U}\mathbf{U}^H = \mathbf{I}$. The matrix **U** is called a *unitary matrix*. The inverse of general $N \times N$ matrix is difficult to calculate and usually obtained using computer software. The inverse of diagonal matrix exists, if none of diagonal entries is equal to zero and can be determined as

$$\mathbf{D} = \begin{bmatrix} d_{11} & 0 & \cdots & 0 \\ 0 & d_{22} & \cdots & 0 \\ \vdots & \vdots & \ddots & \vdots \\ 0 & 0 & \cdots & d_{NN} \end{bmatrix} \Rightarrow \mathbf{D}^{-1} = \begin{bmatrix} 1/d_{11} & 0 & \cdots & 0 \\ 0 & 1/d_{22} & \cdots & 0 \\ \vdots & \vdots & \ddots & \vdots \\ 0 & 0 & \cdots & 1/d_{NN} \end{bmatrix}. \tag{A.3.15}$$

For 2×2 matrix **A**, the inverse is given by

$$\mathbf{A}^{-1} = \begin{bmatrix} a_{11} & a_{12} \\ a_{21} & a_{22} \end{bmatrix}^{-1} = \frac{1}{a_{11}a_{22} - a_{21}a_{12}} \begin{bmatrix} a_{22} & -a_{12} \\ -a_{21} & a_{11} \end{bmatrix}. \tag{A.3.16}$$

For general $N \times M$ matrix **A**, we define the pseudoinverse of the matrix as

$$\mathbf{A}^+ = (\mathbf{A}^H \mathbf{A})^{-1} \mathbf{A}^H. \tag{A.3.17}$$

Note that the product of $\mathbf{A}^+\mathbf{A}$ becomes an $M \times M$ identity matrix $\mathbf{I}_M$.

A.3.3   Matrix decomposition

The eigenvalues of a square matrix $\mathbf{A}$ are scalars, denoted by λ, that satisfy

$$\mathbf{A}\mathbf{x} = \lambda\mathbf{x}, \tag{A.3.18}$$

for certain nonzero vector $\mathbf{x}$. Here, $\mathbf{x}$ is called the eigenvector of $\mathbf{A}$ corresponding to λ. An $N \times N$ matrix will have N eigenvalues. The eigenvalues can be solved from the characteristic equation of $\mathbf{A}$, defined as

$$\det[\mathbf{A} - \lambda\mathbf{I}] = 0. \tag{A.3.19}$$

It can be shown that the determinant of a matrix is equal to the product of its eigenvalues.

A square matrix $\mathbf{A}$ is a Hermitian matrix if its Hermitian is equal to itself, i.e., $\mathbf{A}^H = \mathbf{A}$. The multiplication result of a general matrix and its Hermitian will be a Hermitian matrix, i.e., $(\mathbf{A}\mathbf{A}^H)^H = \mathbf{A}\mathbf{A}^H$. The eigenvalues of a Hermitian matrix are always real. We can decompose a Hermitian matrix $\mathbf{A}$ as

$$\mathbf{A} = \mathbf{U}\Lambda\mathbf{U}^H, \tag{A.3.20}$$

where $\mathbf{U}$ is a unitary matrix, whose columns are the eigenvectors of $\mathbf{A}$, and Λ is a diagonal matrix, whose diagonal entries are the eigenvalues of $\mathbf{A}$, i.e.,

$$\Lambda = \begin{bmatrix} \lambda_1 & 0 & \cdots & 0 \\ 0 & \lambda_2 & \cdots & 0 \\ \vdots & \vdots & \ddots & \vdots \\ 0 & 0 & \cdots & \lambda_N \end{bmatrix}. \tag{A.3.21}$$

Matrix $\mathbf{A}$ is positive definite if all its eigenvalues are positive. Matrix $\mathbf{A}$ is positive semidefinite if none of its eigenvalues is negative. $\mathbf{A}\mathbf{A}^H$ is always positive semidefinite.

A general $N \times M$ matrix $\mathbf{A}$ can be decomposed into

$$\mathbf{A} = \mathbf{U}\Sigma\mathbf{V}^H, \tag{A.3.22}$$

where Σ is diagonal matrix of size $N \times M$, $\mathbf{U}$ of size $N \times N$, and $\mathbf{V}$ of size $M \times M$ are the left and right singular matrix of $\mathbf{A}$, respectively. The diagonal entries of Σ are called the *singular values,* whereas the columns of $\mathbf{V}$ and $\mathbf{U}$ are the right singular vector and right singular vector, respectively, of $\mathbf{A}$. $\mathbf{A}$ has $\min[N, M]$ singular values, where $R_\mathbf{A}$ of them are nonzero. In general, Σ is of the form

$$\Sigma = \begin{bmatrix} \sigma_1 & 0 & \cdots & 0 & 0 & \cdots \\ 0 & \sigma_2 & \cdots & 0 & 0 & \cdots \\ \vdots & \vdots & \ddots & \vdots & \vdots & \cdots \\ 0 & 0 & \cdots & \sigma_{R_\mathbf{A}} & 0 & \cdots \\ 0 & 0 & \cdots & 0 & 0 & \cdots \\ \vdots & \vdots & \vdots & \vdots & \vdots & \ddots \end{bmatrix}_{N \times M}. \tag{A.3.23}$$

Finally, σ_i is related to the ith eigenvalue of $\mathbf{AA}^H$, λ_i, as $\sigma_i = \sqrt{\lambda_i}$. The above decomposition for general matrix $\mathbf{A}$ is the famous SVD.

A.4 Lagrange multipliers

Lagrange multipliers are widely used to find the stationary points of a function of multiple variables subject to one or more constraints [5]. Let $f(\cdot)$ denote a function of D variables, $x_1, x_2, \ldots, x_D$, which can be grouped into a vector $\mathbf{x} = [x_1, x_2, \ldots, x_D]^T$. Consider the basic problem of finding the maximum of $f(\mathbf{x})$ subject to the constraint $g(\mathbf{x}) = 0$, i.e.,

$$\max_{\mathbf{x}} \quad f(\mathbf{x}) \tag{A.4.1}$$

$$\text{s.t. } g(\mathbf{x}) = 0.$$

An elegant way to solve this problem is to form the Lagrangian with the introduction of an undetermined multiplier as

$$J = f(\mathbf{x}) + \lambda g(\mathbf{x}), \tag{A.4.2}$$

where λ is the so-called Lagrange multiplier. The maximum of $f(\mathbf{x})$ subject to the constraint $g(\mathbf{x}) = 0$ can be determined by finding the stationary point of J with respect to both $\mathbf{x}$ and λ. Taking partial derivatives of J with respect to $\mathbf{x}$ and λ and setting them to zero, we arrive at $D + 1$ equations with $D + 1$ unknowns of the stationary point $\mathbf{x}^*$ and λ. Often λ will be eliminated since we are more interested in $\mathbf{x}^*$.

The Lagrange multiplier method can be justified geometrically as following [6]. Note that $\mathbf{x}$ is a D-dimensional vector. The constraint equation $g(\mathbf{x}) = 0$ then specifies a $D - 1$-dimensional surface in the space of $\mathbf{x}$. The stationary point $\mathbf{x}^*$ will reside on this constraint surface. We first note that the gradient of the constraint function $\nabla g(\mathbf{x})$ for any $\mathbf{x}$ on the constraint surface will be orthogonal to the surface. Otherwise, we could increase the value of $g(\mathbf{x})$ by moving a short distance along the surface, which will violate the constraint equation. More formally, we can show this claim with Taylor expansion. Consider a point in the constraint surface that is very close to $\mathbf{x}$, denoted by $\mathbf{x} + \varepsilon$. With Taylor expansion, we have

$$g(\mathbf{x} + \varepsilon) \simeq g(\mathbf{x}) + \nabla g(\mathbf{x})^T \varepsilon. \tag{A.4.3}$$

Since both $\mathbf{x}$ and $\mathbf{x} + \varepsilon$ on the surface, we have $g(\mathbf{x} + \varepsilon) = g(\mathbf{x}) = 0$. It follows that $\nabla g(\mathbf{x})^T \varepsilon \simeq 0$. When $\|\varepsilon\| \to \mathbf{0}$, ε will be parallel to the constraint surface. The gradient vector $\nabla g(\mathbf{x})$ will be orthogonal to the surface.

We also note that the gradient of the objective function $\nabla f(\mathbf{x})$ at the stationary point $\mathbf{x}^*$ will be orthogonal to the constraint surface. Otherwise, we could increase the value of $f(\mathbf{x})$ by moving an incremental distance along the component of $\nabla f(\mathbf{x})$ parallel to the constraint surface, which implies that the original point is no longer stationary. Based on the above observations, we can see that $\nabla f(\mathbf{x})$ and $\nabla g(\mathbf{x})$ at $\mathbf{x}^*$ are parallel vectors. As such, $\mathbf{x}^*$ must satisfy the following equation:

$$\nabla f(\mathbf{x}) + \lambda \nabla g(\mathbf{x}) = 0, \qquad (A.4.4)$$

where $\lambda \neq 0$ is the Lagrange multiplier. We arrive at the same equation by setting the partial derivatives of the Lagrangian in (D.2) with respect to $\mathbf{x}$ to zero.

Bibliography

[1]   A. V. Oppenheim and R. W. Schafer, *Discrete-time Signal Processing*, 3rd ed. Upper Saddle River, NJ: Pearson, 2009.

[2]   A. Papoulis and S. U. Pillai, *Probability, Random Variables, and Stochastic Processes*, New York, NY: McGraw-Hill, 2002.

[3]   B. P. Lathi and Z. Ding, *Modern Digital and Analog Communication Systems*, 4th ed. New York, NY: Oxford University Press, 2010.

[4]   G. Strang, *Linear Algebra and Its Applications*, 4th ed. New York, NY: Academic Press, 2005.

[5]   J. Nocedal and S. J. Wright, *Numerical Optimization*, New York, NY: Springer, 1999.

[6]   C. M. Bishop, *Pattern Recognition and Machine Learning*, New York, NY: Springer, 2005.

Index